ADVANCED ASIC CHIP SYNTHESIS

Using Synopsys® Design Compiler™
and PrimeTime®

Trademark Information

UNIX is a registered trademark of UNIX Systems Laboratories, Inc.
Verilog is a registered trademark of Cadence Design Systems, Inc.
RSPF and DSPF is a trademark of Cadence Design Systems, Inc.
SDF and SPEF is a trademark of Open Verilog International.

Synopsys, PrimeTime, Formality, DesignPower, DesignWare and SOLV-IT! are registered trademarks of Synopsys, Inc.

Design Analyzer, Design Compiler, Test Compiler, VHDL Compiler, HDL Compiler, ECO Compiler, Library Compiler, Synthetic Libraries, DesignTime, Floorplan Manager, characterize, dont_touch, dont_touch_network and uniquify, are trademarks of Synopsys, Inc.

SolvNET is a service mark of Synopsys, Inc.

All other brand or product names mentioned in this document, are trademarks or registered trademarks of their respective companies or organizations.

All ideas and concepts provided in this book are authors own, and are not endorsed by Synopsys, Inc. Synopsys, Inc. is not responsible for information provided in this book.

ADVANCED ASIC CHIP SYNTHESIS

Using Synopsys® Design Compiler™
and PrimeTime®

Himanshu Bhathagar
Conexant Systems, Inc.
(Formerly, Rockwell Semiconductor Systems)

KLUWER ACADEMIC PUBLISHERS
Boston / Dordrecht / London

Distributors for North, Central and South America:
Kluwer Academic Publishers
101 Philip Drive
Assinippi Park
Norwell, Massachusetts 02061 USA
Telephone (781) 871-6600
Fax (781) 871-6528
E-Mail <kluwer@wkap.com>

Distributors for all other countries:
Kluwer Academic Publishers Group
Distribution Centre
Post Office Box 322
3300 AH Dordrecht, THE NETHERLANDS
Telephone 31 78 6392 392
Fax 31 78 6546 474
E-Mail <orderdept@wkap.nl>

 Electronic Services <http://www.wkap.nl>

Library of Congress Cataloging-in-Publication Data

Bhatnagar, Himanshu.
    Advanced ASIC chip synthesis : Using Synopsys Design Compiler and PrimeTime / Himanshu Bhatnagar.
    p. cm.
  ISBN 0-7923-8537-3 (alk. paper)
    1. Application specific integrated circuits--Design and construction--Data processing. 2. Logic design--Data processing. 3. Computer-aided design. 4. Compilers (Computer programs)
I. Title.
TK7874.6.B44   1999
621.39'5--dc21   99-24720
                                                                                      CIP

Copyright © 1999 by Kluwer Academic Publishers

All rights reserved. No part of this publication may be reproduced, stored in a retrieval system or transmitted in any form or by any means, mechanical, photo-copying, recording, or otherwise, without the prior written permission of the publisher, Kluwer Academic Publishers, 101 Philip Drive, Assinippi Park, Norwell, Massachusetts 02061

Printed on acid-free paper.
Printed in the United States of America

*To my wife Nivedita,
and my daughter Nayana*

Contents

Foreword	xv
Preface	xvii
Acknowledgements	xxiii
About The Author	xxv

CHAPTER 1: ASIC DESIGN METHODOLOGY	**1**
1.1 Typical Design Flow	2
1.1.1 Specification and RTL Coding	5
1.1.2 Dynamic Simulation	6
1.1.3 Constraints, Synthesis and Scan Insertion	7
1.1.4 Formal Verification	9
1.1.5 Static Timing Analysis using PrimeTime	10
1.1.6 Placement, Routing and Verification	11
1.1.7 Engineering Change Order	13
1.2 Chapter Summary	14

CHAPTER 2: TUTORIAL		**15**
2.1	Example Design	16
2.2	Initial Setup	17
2.3	Pre-Layout Steps	18
2.3.1	Synthesis	18
2.3.2	Static Timing Analysis using PrimeTime	23
2.3.3	SDF Generation	26
2.3.4	Verification	28
2.4	Floorplanning and Routing	30
2.5	Post-Layout Steps	35
2.5.1	Static Timing Analysis using PrimeTime	36
2.5.2	Post-Layout Optimization	39
2.6	Chapter Summary	42
CHAPTER 3: BASIC CONCEPTS		**43**
3.1	Synopsys Products	43
3.2	Synthesis Environment	45
3.2.1	Startup Files	45
3.2.2	System Library Variables	47
3.3	Objects, Variables and Attributes	48
3.3.1	Design Objects	48
3.3.2	Variables	49
3.3.3	Attributes	51
3.4	Finding Design Objects	52
3.5	Synopsys Formats	53
3.6	Data Organization	54
3.7	Design Entry	55
3.8	Compiler Directives	56
3.8.1	HDL Compiler Directives	57
3.8.2	VHDL Compiler Directives	60
3.9	Chapter Summary	62
CHAPTER 4: SYNOPSYS TECHNOLOGY LIBRARY		**63**
4.1	Library Basics	64
4.1.1	Library Group	64
4.1.2	Library Level Attributes	64

4.1.3	Environment Description	65
4.1.4	Cell Description	70
4.2	Delay Calculation	73
4.2.1	Delay Model	73
4.2.2	Delay Calculation Problems	75
4.3	What is a Good Library?	76
4.4	Chapter Summary	78

CHAPTER 5: PARTITIONING AND CODING STYLES — 79

5.1	Partitioning for Synthesis	80
5.2	What is RTL?	82
5.2.1	Software versus Hardware	82
5.3	General Guidelines	83
5.3.1	Technology Independence	83
5.3.2	Clock Logic	83
5.3.3	No Glue Logic at the Top	84
5.3.4	Module Name Same as File Name	85
5.3.5	Pads Separate from Core Logic	85
5.3.6	Minimize Unnecessary Hierarchy	85
5.3.7	Register All Outputs	85
5.3.8	Guidelines for FSM Synthesis	86
5.4	Logic Inference	86
5.4.1	Incomplete Sensitivity Lists	86
5.4.2	Memory Element Inference	87
5.4.3	Multiplexer Inference	92
5.4.4	Three-State Inference	95
5.5	Order Dependency	96
5.5.1	Blocking versus Non-Blocking Assignments in Verilog	96
5.5.2	Signals versus Variables in VHDL	97
5.6	Chapter Summary	98

CHAPTER 6: CONSTRAINING DESIGNS — 99

6.1	Environment and Constraints	100
6.1.1	Design Environment	100
6.1.2	Design Constraints	105
6.2	Advanced Constraints	110
6.3	Clocking Issues	113

6.3.1		Pre-Layout	114
6.3.2		Post-Layout	115
6.3.3		Generated Clocks	116
6.4		Putting it Together	117
6.5		Chapter Summary	119

CHAPTER 7: OPTIMIZING DESIGNS — 121

7.1		Design Space Exploration	121
7.2		Total Negative Slack	125
7.3		Compilation Strategies	126
7.3.1		Top-Down Hierarchical Compile	127
7.3.2		Time-Budgeting Compile	128
7.3.3		Compile-Characterize-Write-Script-Recompile	130
7.4		Resolving Multiple Instances	134
7.5		Optimization Techniques	135
7.5.1		Compiling the Design	136
7.5.2		Flattening and Structuring	137
7.5.3		Removing Hierarchy	141
7.5.4		Optimizing Clock Networks	142
7.5.5		Optimizing for Area	145
7.6		Chapter Summary	145

CHAPTER 8: DESIGN FOR TEST — 147

8.1		Types of DFT	147
8.1.1		Memory BIST	148
8.1.2		Boundary Scan DFT	148
8.2		Scan Insertion	149
8.2.1		Making Design Scannable	149
8.2.2		Test Pattern Generation	152
8.3		DFT Guidelines	152
8.3.1		Tri-State Bus Contention	152
8.3.2		Latches	153
8.3.3		Gated Reset or Preset	153
8.3.4		Gated or Generated Clocks	153
8.3.5		Use Single Edge of the Clock	154
8.3.6		Multiple Clock Domains	155
8.3.7		Order Scan-Chains to Minimize Clock Skew	155

8.3.8	Logic Un-Scannable due to Memory Element	156
8.4	Chapter Summary	158

CHAPTER 9: LINKS TO LAYOUT & POST-LAYOUT OPT.   159

9.1	Generating Netlist for Layout	160
9.1.1	Uniquify	161
9.1.2	Tailoring the Netlist for Layout	163
9.1.3	Remove Unconnected Ports	164
9.1.4	Visible Port Names	164
9.1.5	Verilog Specific Statements	165
9.1.6	Unintentional Clock or Reset Gating	166
9.1.7	Unresolved References	167
9.2	Layout	167
9.2.1	Floorplanning	167
9.2.2	Clock Tree Insertion	172
9.2.3	Transfer of Clock Tree to Design Compiler	176
9.2.4	Routing	178
9.2.5	Extraction	178
9.3	Post-Layout Optimization	183
9.3.1	Back Annotation and Custom Wire Loads	184
9.3.2	In-Place Optimization	186
9.3.3	Location Based Optimization	187
9.3.4	Fixing Hold-Time Violations	189
9.4	Future Directions	193
9.5	Chapter Summary	194

CHAPTER 10: SDF GENERATION   195

10.1	SDF File	196
10.2	SDF File Generation	198
10.2.1	Generating Pre-Layout SDF File	198
10.2.2	Generating Post-Layout SDF File	201
10.2.3	Issues Related to Timing Checks	202
10.2.4	False Delay Calculation Problem	203
10.2.5	Putting it Together	205
10.3	Chapter Summary	207

CHAPTER 11: PRIMETIME BASICS — 209

11.1	Introduction	210
11.1.1	Invoking PT	210
11.1.2	PrimeTime Environment	210
11.1.3	Automatic Command Conversion	211
11.2	Tcl Basics	212
11.2.1	Command Substitution	213
11.2.2	Lists	213
11.2.3	Flow Control and Loops	215
11.3	PrimeTime Commands	215
11.3.1	Design Entry	215
11.3.2	Clock Specification	216
11.3.3	Timing Analysis Commands	221
11.3.4	Other Miscellaneous Commands	227
11.4	Chapter Summary	230

CHAPTER 12: STATIC TIMING ANALYSIS — 231

12.1	Why Static Timing Analysis?	231
12.1.1	What to Analyze?	232
12.2	Timing Exceptions	233
12.2.1	Multicycle Paths	233
12.2.2	False Paths	237
12.3	Disabling Timing Arcs	239
12.3.1	Disabling Timing Arcs Individually	240
12.3.2	Case Analysis	241
12.4	Environment and Constraints	242
12.4.1	Operating Conditions – A Dilemma	242
12.5	Pre-Layout	243
12.5.1	Pre-Layout Clock Specification	244
12.5.2	Timing Analysis	245
12.6	Post-Layout	247
12.6.1	What to Back Annotate?	248
12.6.2	Post-Layout Clock Specification	249
12.6.3	Timing Analysis	249
12.7	Analyzing Reports	254
12.7.1	Pre-Layout Setup-Time Analysis Report	254
12.7.2	Pre-Layout Hold-Time Analysis Report	256

12.7.3	Post-Layout Setup-Time Analysis Report	258
12.7.4	Post-Layout Hold-Time Analysis Report	260
12.8	Advanced Analysis	262
12.8.1	Detailed Timing Report	262
12.8.2	Cell Swapping	265
12.8.3	Bottleneck Analysis	266
12.8.4	Clock Gating Checks	269
12.9	Chapter Summary	272

APPENDIX   **275**

INDEX   **277**

Foreword

Our semiconductor industry is increasingly characterized by accelerated product obsolescence. As a result, business success is increasingly dependent upon the ability of development teams to deliver a shortest "time-to-market" product that meets customer requirements. Early product introduction means higher profit margins, lasting only until slower-to-market competitors enter and erode prices.

This intense cycle of market price erosion has been particularly evident in the personal computer industry over the last few years. Consumers are continually demanding quality products at lower cost but with increasing features. Semiconductor suppliers are, in turn, driven to develop system-on-a-chip (SoC) products utilizing VDSM (Very-Deep-Sub-Micron) technologies, just to remain competitive.

Several high performance tools and techniques have been developed over the past few years to mitigate somewhat this "time-to-market" pressure and to enable rapid design updates to meet evolving customer specifications. These changes have resulted in a redefinition of standard ASIC design flow methodologies. High level design languages, like VHDL and Verilog, have displaced schematic capture, thus promoting design reuse. Dynamic simulation has given way to formal verification and static timing analysis. In addition, synthesis engines have become more sophisticated, targeting complex designs containing millions of gates and large IP cores. It is now estimated that the number of gates in a complex ASIC will approach 10 million early in the next decade.

Successfully achieving these levels of integration in a time-to-market focused development environment will require an intimate knowledge of ASIC design flow in the VDSM realm and a complex integration of products offered by multiple EDA tool vendors.

This book, written by Himanshu Bhatnagar, provides a comprehensive overview of the ASIC design flow targeted for VDSM technologies using the Synopsys suite of tools. It emphasizes the practical issues faced by the semiconductor design engineer in terms of synthesis and the integration of front-end and back-end tools. Traditional design methodologies are challenged and unique solutions are offered to help define the next-generation of ASIC design flows. The author provides numerous practical examples derived from real-world situations that will prove valuable to practicing ASIC design engineers as well as to students of advanced VLSI courses in ASIC design.

Dr. Dwight W. Decker
Chairman and CEO, Conexant Systems, Inc.
(Formerly, Rockwell Semiconductor Systems)
Newport Beach, California, U.S.A.

Preface

This book describes the advanced concepts and techniques used towards ASIC chip synthesis, formal verification and static timing analysis, using the Synopsys suite of tools. In addition, the entire ASIC design flow methodology targeted for VDSM (Very-Deep-Sub-Micron) technologies is covered in detail.

The emphasis of this book is on real-time application of Synopsys tools, used to combat various problems seen at VDSM geometries. Readers will be exposed to an effective design methodology for handling complex, sub-micron ASIC designs. Significance is placed on HDL coding styles, synthesis and optimization, dynamic simulation, formal verification, DFT scan insertion, links to layout, and static timing analysis. At each step, problems related to each phase of the design flow are identified, with solutions and work-arounds described in detail. In addition, crucial issues related to layout, which includes clock tree synthesis and back-end integration (links to layout) are also discussed at length. Furthermore, the book contains in-depth discussions on the basics of Synopsys technology libraries and HDL coding styles, targeted towards optimal synthesis solution.

Target audiences for this book are practicing ASIC design engineers and graduate level students undertaking advanced VLSI courses on ASIC chip design and DFT techniques.

This book is not intended as a substitute or a replacement for the Synopsys reference manual, but is meant for anyone who is involved in the ASIC design flow. Also, it is useful for those designers (and companies) who do not have layout capability, or their own technology libraries, but rely on outside vendors for back-end integration and final fabrication of the device. The book provides alternatives to traditional method of netlist hand-off to outside vendors because of various issues related to VDSM technologies. It also addresses solutions to common problems faced by designers when interfacing various tools from different EDA tool vendors.

Overview of the Chapters

Chapter 1 presents an overview to various stages involved in the ASIC design flow using Synopsys tools. The entire design flow is briefly described, starting from concept to chip tape-out. This chapter is useful for designers who have not delved in the full process of chip design and integration, but would like to learn the full process of ASIC design flow.

Chapter 2, outlines the practical aspects of the ASIC design flow as described in Chapter 1. Beginners may use this chapter as a tutorial. Advanced users of Synopsys tools may benefit by using this chapter as a reference. Users with no prior experience in synthesis using Synopsys tools should skip this chapter and return to it later after reading the remaining book.

The basic concepts related to synthesis are described in detail in Chapter 3. These concepts introduce the reader to synthesis terminology used throughout the later chapters. Readers will find the information provided here useful by gaining a basic understanding of these tools and their environment. In addition to describing the purpose of each tool and their setup, this chapter also focuses on defining objects, variables, attributes and compiler directives used by the Design Compiler.

Chapter 4 describes the basics of the Synopsys technology library. Designers usually do not concern themselves with the full details of the technology library, as long as the library contains a variety of cells with different drive strengths. However, a rich library usually determines the quality of synthesis. Therefore, the intent of this chapter is to describe the Synopsys technology library from the designer's perspective. Focus is provided on delay calculation method and other techniques that designers may use in order to alter the behavior of the technology library, hence the quality of the synthesized design.

Proper partitioning and good coding style is essential in obtaining quality results. Chapter 5 provides guidelines to various techniques that may be used to correctly partition the design in order to achieve the optimal solution. In addition, the HDL coding styles is covered in this chapter that illustrates numerous examples and provides recommendations to designers on how to code the design in order to produce faster logic and minimum area.

The Design Compiler commands used for synthesis and optimization are described in Chapter 6. This chapter contains information that is useful for the novice and the advanced users of Synopsys tools. The chapter focuses on real-world applications by taking into account deviations from the ideal situation i.e., "Not all designs or designers, follow Synopsys recommendations". The chapter illustrates numerous examples that help guide the user in real-time application of the commands.

Chapter 7 discusses optimization techniques in order to meet timing and area requirements. Comparison between older version of Design Compiler and the new version is highlighted. Emphasis is provided on the new optimization technique employed by Design Compiler called "TNS". Also, detailed analysis on various methods used for optimizing logic is presented. In addition, different compilation strategies, each with advantages and disadvantages are discussed in detail.

DFT techniques are increasingly gaining momentum among ASIC design engineers. Chapter 8 provides a brief overview of the different types of DFT techniques that are in use today, followed by detailed description on how devices can be made scannable using Synopsys's Test Compiler. It describes commands used for inserting scan through Design Compiler. A multitude of guidelines is presented in order to alleviate the problems related to DFT scan insertion on a design.

Chapter 9 discusses the links to layout feature of Design Compiler. It describes the interface between the front-end and back-end tools. Also, this chapter provides different strategies used for post-layout optimization of designs. This includes in-place and location based optimization techniques. Furthermore, a section is devoted to clock tree insertion and issues related to clock tree transfer to Design Compiler. Various solutions to this common problem are described. This chapter is extremely valuable for designers (and companies) who do not posses their own layout tool, but would like to learn the place and route process along with full chip integration techniques.

Chapter 10, titled "SDF Generation: for Dynamic Timing Simulation" describes the process of generating the SDF file from Design Compiler or PrimeTime. A section is devoted to the syntax of SDF format, followed by detailed discussion on the process of SDF generation, both for pre and post-layout phases of the design. In addition, few innovative ideas and suggestions are provided to facilitate designers in performing successful simulation. This chapter is useful for those designers who prefer dynamic simulation method to formal verification techniques, in order to verify the functionality of the design.

Chapter 11 introduces to the reader, the basics of static timing analysis, using PrimeTime. This includes a brief section devoted to Tcl language that is utilized by PrimeTime. Also described in this chapter are selected PrimeTime commands that are used to perform static timing analysis, and also facilitate the designer in debugging the design for possible timing violations.

The key to working silicon usually lies in successful completion of static timing analysis performed on a particular design. This capability makes static timing analysis one of the most important steps in the entire design flow and is used by many designers as a sign-off criterion to the ASIC vendor. Chapter 12 is devoted to several basic and advanced topics on static timing analysis, using PrimeTime. It effectively illustrates the usage of PrimeTime, both for the pre and the post-layout phases of the ASIC design flow process. In addition, numerous examples on analyzing reports and suggestions on various scenarios are provided. This chapter is useful to those who would like to migrate from traditional methods of dynamic simulation to the method of analyzing designs statically. It is also helpful for those readers who would like to perform in-depth analysis of the design through PrimeTime.

Conventions Used in the Book

All Synopsys commands are typed in "Ariel" font. This includes all examples that contain synthesis and timing analysis scripts.

The command line prompt is typed in "`Courier New`" font. For example:

`dc_shell>` and, `pt_shell>`

Option values for some of the commands are enclosed in < and >. In general, these values need to be replaced before the command can be used. For example:

set_false_path –from <from list> –to <to list>

The "\" character is used to denote line continuation, whereas the "|" character represents the "OR" function. For example:

compile –map_effort low | medium | high \
   –incremental_mapping

Wherever possible, keywords are *italicized*. Topics or points, that need emphasis are underlined or highlighted through **bold** font.

Acknowledgements

I would like to express my heartfelt gratitude to a number of people who contributed their time and effort towards this book. Without their help, it would have been impossible to take this enormous undertaking.

First and foremost, a special thanks to my family, who gave me continuous support and encouragement that kept me constantly motivated towards the completion of this project. My wife Nivedita, who patiently withstood my nocturnal and weekend writing activities, spent enormous amount of time towards proofreading the manuscript and correcting my "Engineers English". I could not have accomplished this task without her help and understanding.

I would like to thank my supervisor, Anil Mankar for giving me ample latitude at work, to write the book. His moral support and innovative suggestions kept me alert and hopeful. I would also like to thank my colleagues at Conexant; Hoat Nguyen, Dao Doan, Chilan Nguyen, Randy Kolar, Chung Jue Chen, Chih-Shun Ding, Steve Schulz, Khosrow Golshan, Richard Ward, Sameer Rao and Ravi Ranjan who devoted their precious time in reviewing the manuscript.

I was extremely fortunate to have an outstanding reviewer for this project, Dr. Kelvin F. Poole (Clemson University, S.C.). I have known Dr. Poole for a number of years and approached him for his guidance while writing this book. He not only proofread the entire manuscript word-by-word (gritting his teeth, I'm sure!), but also provided valuable suggestions, which helped make the book more robust. Thank you Dr. Poole.

I wish to express my thanks to Bill Mullen, Ahsan Bootehsaz, Steve Meier, Russ Segal, Juergen Froessl and Amanda Hsiao at Synopsys, who participated in reviewing this manuscript and provided me with many valuable suggestions. Julie Liedtke and Bryn Ekroot of Synopsys helped me write the necessary Trademark information. Special thanks are also due to Kameshwar Rao, Jeff Echtenkamp, Heratch Avakian, and Chin-Sieh Lee of Broadcom Corporation for providing me valuable feedback and engaging in lengthy technical discussions. Thanks are also due to Jean-Claude Marin (SGS-Thomson Microelectronics, France), Tapan Mohanti (Fairchild Semiconductors), Dr. Sudhir Aggarwal (Philips Semiconductors), Abu Horaira (Intel Corporation), Phong Tran (Chameleon Technologies), Karim Hussain, Kevin Walis and Ginsy Chagger (Factor Technologies, U.K.) for giving me positive feedback at all times. Their endless encouragement is very much appreciated.

When I first started this project, I contacted Carl Harris of Kluwer Academic Publishers and described to him this idea. He immediately got excited and gave me the go-ahead. Although I taxed his patience many times, he still kept me on my toes and pushed me towards completing this book. His understanding even when I kept on delaying the book is appreciated.

A final word, "Thank you Mom and Dad for all the sacrifices you made to further my career".

Himanshu Bhatnagar
Conexant Systems, Inc.
Newport Beach, California

About The Author

Himanshu Bhatnagar is a Senior ASIC Design Engineer at Conexant Systems, Inc. based in Newport Beach, California. Conexant Systems Inc., formerly Rockwell Semiconductor Systems, is the world's largest independent company focused exclusively on providing semiconductor products for communication electronics. Himanshu has been instrumental in defining the next generation ASIC design flow methodologies using latest high performance tools from Synopsys and other EDA tool vendors.

Before Joining Conexant, Himanshu worked for SGS-Thomson Microelectronics in Singapore and the corporate headquarters based in Grenoble, France. He completed his undergraduate degree in Electronics and Computer Science from Swansea University (Wales, U.K), and his masters degree in VLSI design from Clemson University, (South Carolina, USA).

1

ASIC DESIGN METHODOLOGY

As deep sub-micron semiconductor geometries shrink, traditional methods of chip design have become increasingly difficult. In addition, increasing numbers of transistors are being packed into the same die-size that makes it extremely hard, if not impossible to validate the design. Furthermore, the ever critical "time-to-market" the chip has remained the same, or is under constant pressure to be reduced. To counteract these problems, new methods and tools have evolved to facilitate the ASIC design methodology.

The main function of this chapter is bringing to the forefront different stages involved in chip design as we move deeper into the sub-micron realm. Various techniques that improve the design flow are also discussed.

This chapter focuses on the entire synthesis based ASIC design flow methodology, using Synopsys tools, from RTL coding to the final tape-out. Also discussed are various aspects of layout related issues.

Also, it is worth mentioning that the concept of formal verification described in this chapter is new to the IC design world, and as yet, not embraced one hundred percent by the ASIC design community. This chapter will attempt to stress the importance of this technique (using Synopsys Formality tool) to the reader, and explain the necessity of this technique in the design flow to achieve the maximum benefit, by reducing the overall cycle time.

1.1 Typical Design Flow

The ASIC design flow contains the steps outlined below. Figure 1-1 illustrates the flow chart relating to the design flow described below. Synthesis related topics are described in detail in subsequent chapters.

1. Architectural and electrical specification.

2. RTL coding and simulation test bench preparation.

3. DFT memory BIST insertion, for designs containing memory elements.

4. Exhaustive dynamic simulation of the design, in order to verify the functionality of the design.

5. Design environment setting. This includes the technology library to be used, along with other environmental attributes.

6. Constraining and synthesizing the design with scan insertion (and optional JTAG) using Design Compiler.

7. Block level static timing analysis, using Design Compiler's built-in static timing analysis engine.

8. Formal verification of the design. RTL against the synthesized netlist, using Formality.

9. Pre-layout static timing analysis on the full design through PrimeTime.

10. Forward annotation of timing constraints to the layout tool.

11. Initial floorplanning with timing driven placement of cells, clock tree insertion and global routing

12. Transfer of clock tree to the original design (netlist) residing in Design Compiler.

13. Formal verification between the synthesized netlist and clock tree inserted netlist, using Formality.

14. Extraction of estimated timing delays from the layout after the global routing step (step 11).

15. Back annotation of estimated timing data from the global routed design, to Design Compiler and PrimeTime.

16. Static timing analysis in PrimeTime, using the estimated delays extracted after performing global route.

17. In-place optimization of the design in Design Compiler.

18. Detailed routing of the design.

19. Extraction of real timing delays from the detailed routed design.

20. Back annotation of the real extracted data from the detailed routed design, to Design Compiler and PrimeTime.

21. Post-layout static timing analysis using PrimeTime.

22. In-place optimization of the design in Design Compiler, if necessary.

23. Functional gate-level simulation of the design with post-layout timing (if desired).

24. Tape out after LVS and DRC verification.

Figure 1-1, graphically illustrates the typical ASIC design flow discussed above. The acronym STA and CT, represent static timing analysis and clock tree respectively. DC symbolizes Design Compiler.

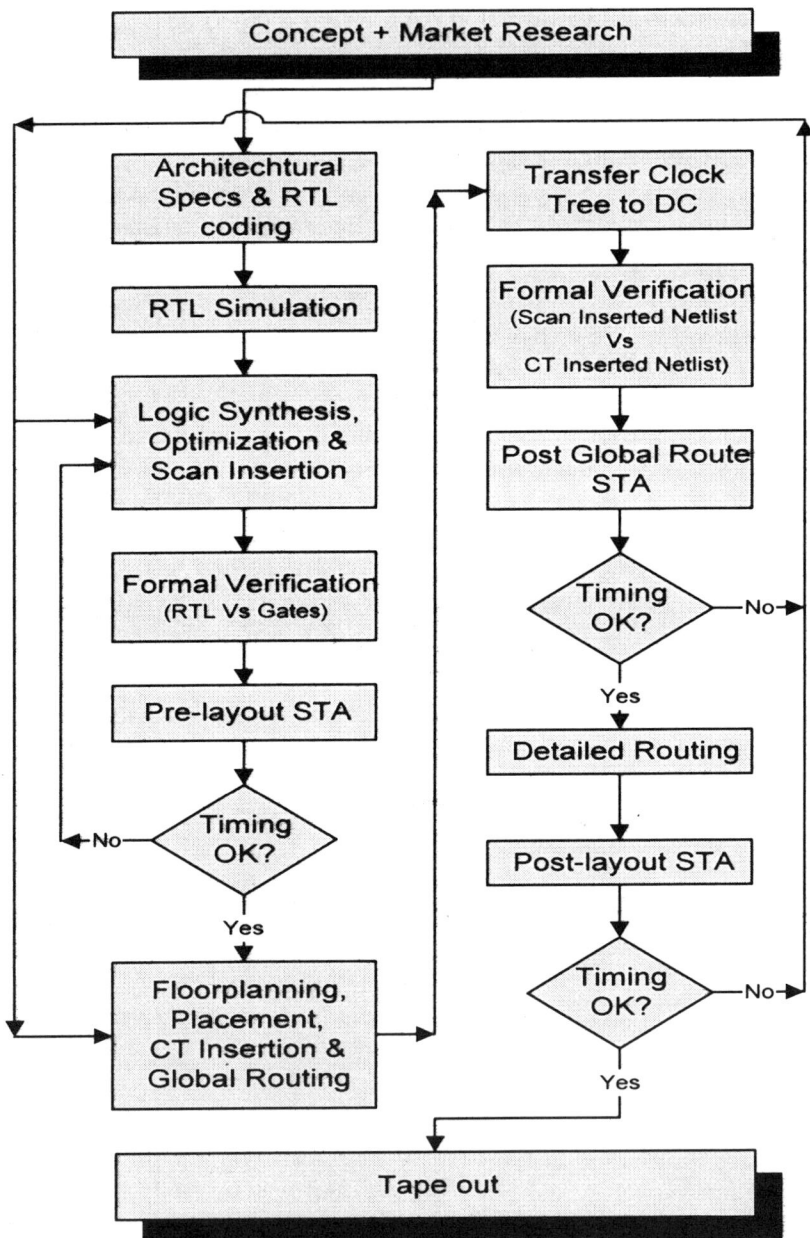

Figure 1-1. ASIC Design Flow

1.1.1 Specification and RTL Coding

As with any product, chip design commences with the conception of an idea dictated by the market. These ideas are then translated into architectural and electrical specifications. The architectural specifications define the functionality and partitioning of the chip into several manageable blocks, while the electrical specifications define the relationship between the blocks in terms of timing information.

The next phase involves the implementation of these specifications. In the past this was achieved by manually drawing the schematics, utilizing the components of a cell library. This process was time consuming and was impractical for design reuse. To overcome this problem, hardware description languages (HDL) were developed. As the name suggests, the functionality of the design is coded using the HDL. There are two main HDLs in use today, Verilog and VHDL. Both languages perform the same function, each having their own advantages and disadvantages.

There are three levels of abstraction that may be used to represent the design; Behavioral, RTL (Register Transfer Level) and Structural. The Behavioral level code is at a higher level of abstraction. It is used primarily for translating the architectural specification, to a code that can be simulated. Behavioral coding is initially performed to explore the authenticity and feasibility of the chosen implementation for the design. Conversely, the RTL coding actually describes and infers the structural components and their connections. This type of coding is used to describe the functionality of the design and is synthesizable to produce the structural netlist, which uses the leaf cells of a library.

The design is coded using the RTL style, in either Verilog or VHDL, or both. It can also be partitioned if necessary, into a number of smaller blocks to form a hierarchy, with a top-level block connecting all lower level blocks.

☞ Synopsys recently introduced Behavior Compiler, capable of synthesizing Behavior level style of coding. Since this is a major topic of discussion and is not relevant to this book, only RTL related synthesis is covered in this book.

1.1.2 Dynamic Simulation

The next step is to check the functionality of the design by simulating the RTL code. All currently available simulators are capable of simulating the behavior level as well as RTL level coding styles. In addition, they are also used to simulate the mapped gate-level design.

Figure 1-2, illustrates a partitioned design surrounded by a test bench ready for simulation. This test bench is normally written in behavior HDL while the actual design is coded in RTL.

Usually the simulators are language dependent (either Verilog or VHDL), although there are a few simulators in the market, capable of simulating a mixed HDL design.

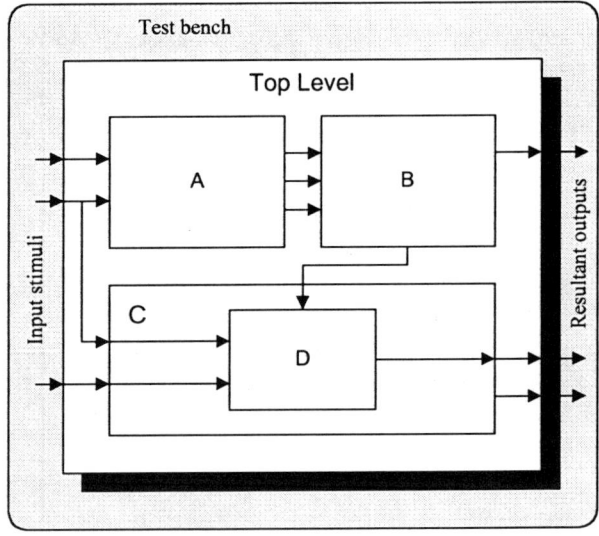

Figure 1-2. Design Hierarchy Example

The purpose of the test bench is to provide necessary stimuli to the design. It is important to note that the coverage of the design is totally dependent on the number of tests performed and the quality of the test bench. This is the reason why a sound test bench is extremely critical to the design. During the simulation of the RTL, the component (or gate) timing is not considered. Therefore, to minimize the difference between the RTL simulation and the synthesized gate-level simulation at a later stage, the delays are usually coded within the RTL source, usually for sequential elements.

1.1.3 Constraints, Synthesis and Scan Insertion

For a long time, the HDLs were used for logic verification. Designers would manually translate the HDL into schematics and draw the interconnections between the components to produce a gate-level netlist. With the advent of synthesis tools, this manual task has been rendered obsolete. The tool has taken over and performs the task of reducing the RTL to the gate-level netlist. This process is termed as synthesis.

Synopsys's Design Compiler (from now on termed as, DC) is the de-facto standard and by far the most popular synthesis tool in the ASIC industry today.

Synthesizing a design is an iterative process and begins with defining timing constraints for each block of the design. These timing constraints define the relationship of each signal with respect to the clock input for a particular block. In addition to the constraints, a file defining the synthesis environment is also needed. The environment file specifies the technology cell libraries and other relevant information that DC uses during synthesis.

DC reads the RTL code of the design and using the timing constraints, synthesizes the code to structural level, thereby producing a mapped gate-level netlist. This concept is shown in Figure 1-3.

Usually, for small blocks of a design, DC's internal static timing analysis is used for reporting the timing information of the synthesized design. DC tries to optimize the design to meet the specified timing constraints. Further steps may be necessary if timing requirements are not met.

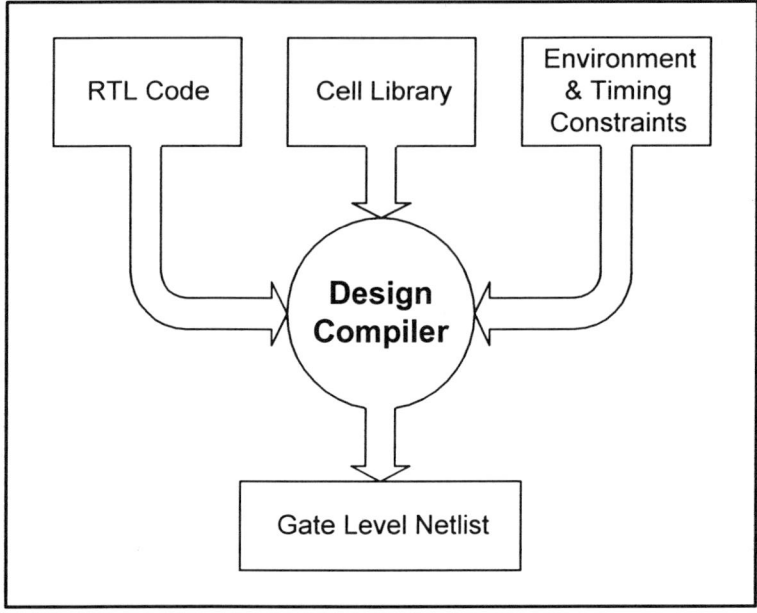

Figure 1-3. Design Compiler Inputs and Outputs

Most designs today, incorporate design-for-test (DFT) logic to test their functionality, after the chip is fabricated. The DFT consists of memory BIST (built-in-self-test), scan insertion and Boundary Scan (JTAG) etc.

The memory BIST comprises of synthesizable RTL that is based upon controller logic and is incorporated in the design before synthesis. There are tools available in the market that may be used to generate the BIST controller and surrounding logic. Unfortunately, Synopsys does not provide this capability.

The scan insertion may be performed using the test ready compile feature of DC. This procedure maps the RTL directly to scan-flops, before linking them in a scan-chain. An advantage of using this feature is its ability to enable DC to take the scan-flop timing into account while synthesizing. This technique is important since the scan-flops generally have different delays associated with them as compared to their non-scan equivalent flops (or normal flops).

JTAG or boundary scan is primarily used for testing the board connections, without unplugging the chip from the board. The JTAG controller and surrounding logic may also be generated directly by DC.

A number of steps must be performed in order to perform successful synthesis. These will be discussed later in subsequent chapters. For the moment, however, the process illustrated above is sufficient for the purpose of explaining the design flow.

1.1.4 Formal Verification

The concept of formal verification is fairly new to the ASIC design community. Formal verification techniques perform validation of a design using mathematical methods without the need for technological considerations, such as timing and physical effects. They check for logical functions of a design by comparing it against the reference design.

A number of EDA tool vendors have developed the formal verification tools. However, only recently, Synopsys also introduced to the market its own formal verification tool called Formality.

The main difference between formal methods and dynamic simulation is that the former technique verifies the design by proving that the structure and functionality of two designs are logically equivalent. Dynamic simulation methods can only probe certain paths of the design that are sensitized, thus may not catch a problem present elsewhere. In addition, formal methods consume negligible amount of time as compared to dynamic simulation.

The purpose of the formal verification in the design flow is to validate the RTL against RTL, gate-level netlist against the RTL code, or the comparison between gate-level to gate-level netlists.

The RTL to RTL verification is used to validate the new RTL against the old functionally correct RTL. This is usually performed for designs that are subject to frequent changes in order to accommodate additional features. When these features are added to the source RTL, there is always a risk of breaking the old functionally correct feature. To prevent this, formal verification may be performed between the old RTL and the new RTL to check the validity of the old functionality.

The RTL to gate-level verification is used to ascertain that the logic has been synthesized accurately by DC. Since the RTL is dynamically simulated to be functionally correct, the formal verification of the design between the RTL and the scan inserted gate-level netlist assures us that the gate-level also has the same functionality. In this instance if we were to use the dynamic simulation method to verify the gate-level, it would have taken a long time (days and weeks, depending on the size of the design) to verify the design. In comparison, the formal method would take a few hours to perform a similar verification.

The last part involves verifying the gate-level netlist against the gate-level netlist. This too is a significant step for the verification process, since it is mainly used to verify – what has gone into the layout versus what has come out of the layout. What comes out of the layout is obviously the clock tree inserted netlist (flat or hierarchical). This means that the original netlist that goes into the layout tool is modified. The formal technique is used to verify the logic equivalency of the modified netlist against the original netlist.

1.1.5 Static Timing Analysis using PrimeTime

As previously mentioned, the block level static timing analysis is done using DC. Although, the chip-level static timing can be performed using the above approach, it is recommended that PrimeTime, be used instead. PrimeTime is the Synopsys stand-alone sign-off quality static timing analysis tool that is capable of performing extremely fast static timing analysis on full chip-level designs. It provides a Tcl interface that provides a powerful environment for analysis and debugging of designs.

The static timing analysis, to some extent, is the most important step in the whole ASIC design process. This analysis allows the user to exhaustively analyze all critical paths of the design and express it in an orderly report.

Furthermore, the report can also contain other debugging information like the fanout or capacitive loading of each net.

The static timing is performed both for the pre and post-layout gate-level netlist. In the pre-layout mode, PrimeTime uses the wire load models specified in the library to estimate the net delays. During this, the same timing constraints that were fed to DC previously are also fed to PrimeTime, specifying the relationship between the primary I/O signals and the clock. If the timing for all critical paths is acceptable, then a constraints file may be written out from PrimeTime or DC for the purpose of forward annotation to the layout tool. This constraint file in SDF format specifies the timing between each group of logic that the layout tool uses, in order to perform the timing driven placement of cells.

In the post-layout mode, the actual extracted delays are back annotated to PrimeTime to provide realistic delay calculation. These delays consist of the net capacitances and interconnect RC delays.

Similar to synthesis, static timing analysis is also an iterative process. It is closely linked with the placement and routing of the chip. This operation is usually performed a number of times until the timing requirements are satisfied.

1.1.6 Placement, Routing and Verification

As the name suggests, the layout tool performs the placement and routing. There are a number of methods in which this step could be performed. However, only issues related to synthesis are discussed in this section.

The quality of floorplan and placement is more critical than the actual routing. Optimal cell placement location, not only speeds up the final routing, but also produces superior results in terms of timing and reduced congestion. As explained previously, the constraint file is used to perform timing driven placement. Although this is the recommended approach, sometimes this approach leads to significant impact on the overall area. However, in general, the area is decreased due to rubber-banding effect of the timing driven placement approach. It is up to the user's discretion to decide whether timing driven placement is a valid approach. The timing driven placement method forces the layout tool to place the cells according to the criticality of the timing between the cells.

After the placement of cells, the clock tree is inserted in the design by the layout tool. The clock tree insertion is optional and depends solely on the design and user's preference. Users may opt to use more traditional methods of routing the clock network, for example, using fishbone/spine structure for the clocks in order to reduce the total delay and skew of the clock. As technologies shrink, the spine approach is getting more difficult to implement due to the increase in resistance (thus, RC delays) of the interconnect wires. It is therefore the intent of this section (and the entire book) to stress solely on the clock tree synthesis approach.

At this stage an additional step is necessary to complete the clock tree insertion. As mentioned above, the layout tool inserted the clock tree in the design after the placement of cells. Therefore, the original netlist that was generated from DC (and fed to the layout tool), lacks the clock tree information (essentially the whole clock tree network, including buffers and nets). Therefore, the clock tree must be re-inserted in the original netlist and formally verified. Some layout tools provide direct interface to DC to perform this step. Chapter 9 introduces some of these steps, both traditional and not-so-traditional approaches. For the sake of simplicity, lets assume that the clock tree insertion to the original netlist has been performed.

The layout tool generally performs routing in two phases – global routing and detailed routing. After placement, the design is globally routed to determine the quality of placement, and to provide estimated delays approximating the real delay values of the post-routed (after detailed routing) design. If the cell placement is not optimal, the global routing will take a longer time to complete, as compared to placing the cells. Bad placement also affects the overall timing of the design. Therefore, to minimize the number of synthesis-layout iterations and improve placement quality, the timing information is extracted from the layout, after the global routing phase. Although, these delay numbers are not as accurate as the numbers extracted after detailed routing, they do provide a fair idea of the post-routed timing. The estimated delays are back annotated to PrimeTime for analysis,

ASIC DESIGN METHODOLOGY 13

and only when the timing is considered satisfactory, the remaining process is allowed to proceed.

Detailed routing is the final step that is performed by the layout tool. After detailed route is complete, the real timing delays of the chip are extracted, and plugged into PrimeTime for analysis.

These steps are iterative and depend on the timing margins of the design. If the design fails timing requirements, post-layout optimization is performed on the design before undergoing another iteration of layout. If the design passes static timing analysis, it is ready to undergo LVS (layout versus schematic) and DRC (design rule checking) before tape-out.

It must be noted that all steps discussed above can also be applied for hierarchical place and route. In other words, one can repeat these steps for each sub-block of the design before placing the sub-blocks together in the final layout and routing between the sub-blocks.

1.1.7 Engineering Change Order

This step is an exception to the normal design flow and should not be confused with the regular design cycle. Therefore, this step will not be explained in subsequent chapters.

Many designers regard engineering change order (ECO) as the change required in the netlist at the very last stage of the ASIC design flow. For instance, ECO is performed when there is a hardware bug encountered in the design at the very last stage (say, after tape-out), and it is necessary to perform a metal mask change by re-routing a small portion of the design.

As a result ECO is performed on a small portion of the chip to prevent disturbing the placement and routing of the rest of the chip, thereby preserving the rest of the chip's timing. Only the part that is affected is modified. This can be achieved, either by targeting the spare gates incorporated in the chip, or by routing only some of the metal layers. This process is termed as metal mask change.

Normally, this procedure is executed for changes that require less than 10% modification of the whole chip (or a block, if doing hierarchical place and route). If the bug fix requires more than 10% change then it is best to repeat the whole procedure and re-route the chip (or the block).

The latest version of DC incorporates the ECO compiler. It makes use of the mathematical algorithms (also used by the formal verification techniques), to automatically implement the required changes. Making use of the ECO compiler provides designers an alternative to the tedium of manually inserting the required changes in the netlist, thus minimizing the turn-around time of the chip.

Some layout tools have incorporated the ECO algorithm within their tool. The layout tool has a built-in advantage that it does not suffer from the limitation of crossing the hierarchical boundaries associated with a design. Also, the layout tool benefits from knowing the placement location of the spare cells (normally included by the designers in the design), thus can target the nearest location of spare cells in order to implement the required ECO changes and achieve minimized routing.

1.2 Chapter Summary

In this chapter the ASIC design flow incorporating the latest tools and technology for very deep sub-micron (VDSM) technologies were reviewed. The flow started with the definition of specification, and ended with physical layout. The significance was placed on logic synthesis related topics and clock tree insertion.

Also introduced was a new concept of formal verification as applicable to the design flow to shorten the design cycle of the chip. The need to perform formal verification was emphasized to eliminate the need for dynamic gate-level simulation.

Finally, a section was devoted to performing ECO on the design. Different concepts were introduced that may make it easier for the reader to perform this step.

2

TUTORIAL

Synthesis and Static Timing Analysis

This chapter is intended both for beginners and advanced users of Synopsys tools. Novices with no prior experience in synthesis using Synopsys tools are advised to skip this chapter and return to it after reading the remainder of the book. Beginners with minimal experience in synthesis may use this chapter as a jump-start to learn the ASIC design process, using Synopsys tools. Advanced users will benefit by using this chapter as a reference.

The chapter offers minimal or no explanation, for Synopsys commands (they are explained in subsequent chapters). The emphasis is on outlining the practical aspects of the ASIC design flow described in Chapter 1, with Synopsys synthesis in the center. This helps the reader correlate the theoretical concepts with its practical application.

Although, the previous chapter stressed skipping the gate-level simulation in favor of formal verification techniques, many designers are reluctant to forego the former step. Due to this reason, this chapter also covers the SDF generation from DC, to be used for simulation purposes. Also, the chapter includes static timing analysis using PrimeTime (PT), in addition to application of formal verification methods, using Formality.

Synthesis and optimization may be performed using any number of approaches. This solely depends upon the methodology you prefer, or are most comfortable using. This chapter uses one such approach that is most commonly used by the Synopsys user's community. You may cater this approach to suit your individual requirements with relative ease.

For the sake of clarity and ease of explanation, the bottom-up compile methodology (described later) is used in all examples and scripts, relating to the synthesis process presented in this chapter. Also, it must be noted that the entire ASIC flow is extremely iterative and one should not assume that the process described in this chapter is suitable for all designs. Later chapters discuss each topic in detail that can be tailored to your designs and methodology.

2.1 Example Design

The best way to start this topic is to go through the whole process on an example design. A tap controller design, coded in Verilog HDL and consisting of one level of hierarchy as shown below is chosen for this purpose:

> *tap_controller.v*
>   *tap_bypass.v*
>   *tap_instruction.v*
>   *tap_state.v*

The top level of the design is called *tap_controller* which instantiates three modules called *tap_bypass*, *tap_instruction* and *tap_state*. This design contains a single 30 MHz clock called "tck" and a reset called "trst". Timing specifications for this design dictate that the setup-time needed for all input signals with respect to "tck" is 10ns, while the hold-time is 0ns. Furthermore, all output signals must be delayed by 10ns with respect to the clock.

The process technology targeted for this design is 0.25 micron. In order to achieve greater accuracy due of variance in process, two Synopsys standard cell technology libraries, characterized for worst-case and the best-case process parameters are used. The libraries are called *ex25_worst.db* and *ex25_best.db*, with a corresponding symbol library containing schematic representations, called *ex25.sdb*. The name of the operating conditions defined in the *ex25_worst.db* library is WORST, while the name of the operating conditions in the *ex25_best.db* library is BEST.

It is assumed that the functionality of the design has been verified by dynamically simulating it at the RTL level.

2.2 Initial Setup

The next step is to synthesize the design, i.e., map the design to the gates belonging to the specified technology library. Before we begin synthesis, several setup files must be created as follows:

a) .synopsys_dc.setup file for DC.
b) .synopsys_pt.setup file for PT.

The first file is the setup file for DC and is used for synthesis, while the second file is associated with PT and defines the required setup to be used for static timing analysis.

Create both of these files with the following contents, assuming that the libraries are kept in the directory – /usr/golden/library/std_cells/

DC .synopsys_dc.setup file

```
search_path = search_path + {  "."    "/usr/golden/library/std_cells" }

target_library    = {ex25_worst.db}
link_library      = {"*", ex25_worst.db ex25_best.db}
symbol_library    = {ex25.sdb}

define_name_rules BORG –allowed "A-Za-z0-9_"                    \
                –first_restricted "_" –last_restricted "_"      \
                –max_length 30                                  \
                –map { {"\*cell\*", "mycell"}, {"*–return", "myreturn"} }
```

```
bus_naming_style                      = %s[%d]
verilogout_no_tri                     = true
verilogout_show_unconnected_pins      = true
test_default_scan_style               = multiplexed_flip_flop
```

PT .synopsys_pt.setup file

```
set search_path  [list . /usr/golden/library/std_cells]
set link_path    [list {*} ex25_worst.db, ex25_best.db]
```

2.3 Pre-Layout Steps

The following sub-sections illustrate the steps involved during the pre-layout phase. This includes one-pass logic synthesis with scan insertion, static timing analysis, SDF generation to perform functional gate-level simulation, and finally formal verification between the source RTL and synthesized netlist.

2.3.1 Synthesis

The pre-layout logic synthesis involves optimizing the design for maximum setup-time, utilizing the statistical wire-load models and the worst-case operating conditions from the *ex25_worst.db* technology library. In order to maximize the setup-time, you may constrain the design by defining clock uncertainty for the setup-time. In general, a 10% over-constrain is usually sufficient, in order to minimize the synthesis-layout iterations.

After initial synthesis if gross hold-time violations are detected, they should be fixed at the pre-layout level. This also helps in reducing the synthesis-layout iterations. However, it is preferable to fix minor hold-time violations after the layout, with real delays back annotated.

In this tutorial, we assume that minor hold-time violations exist. Therefore these violations will be fixed during the post-layout optimization. Fixing hold-time violations involves back annotation of the extracted delays from the layout to DC. In addition, hold-time fixes require usage of the best-case operating conditions from the *ex25_best.db* library.

Generic synthesis script for sub-modules

```
/**********************************************/
/* Define design name */
    active_design = tap_bypass

/**********************************************/
/* Design entry in Verilog format */
    analyze –format verilog active_design + ".v"
    elaborate active_design

    current_design active_design
    link

    uniquify

/**********************************************/
/* Design environment */
    set_wire_load  SMALL  –mode top
    set_operating_conditions  WORST

/**********************************************/
/* Clock specification and design constraints */
    create_clock  –period 33  –waveform { 0 16.5}  tck
    set_dont_touch_network  {tck  trst}
    set_clock_skew  –delay 2.0  –minus_uncertainty 3.0  tck

    set_driving_cell  –cell BUFF1X  –pin Z  all_inputs()
    set_drive 0 {tck trst}

    set_input_delay    20.0  –clock tck  –max all_inputs()
    set_output_delay  10.0  –clock tck  –max all_outputs()

    set_max_area 0
```

```
/*************************************************/
/* Compilation with scan insertion */
    set_fix_multiple_port_nets –buffer_constants –all
    compile –scan

    check_test
    create_test_patterns –sample 10
    preview_scan
    insert_scan
    check_test

/*************************************************/
/* Clean-up and generate the netlist in verilog and db formats */
    remove_unconnected_ports find(–hierarchy cell, "*")

    change_names –h –rules BORG

    set_dont_touch current_design

    write –hierarchy –output active_design + ".db"
    write –format verilog –hierarchy         \
                        –output active_design + ".sv"
```

The above script contains a user-defined variable called *active_design* that defines the name of the module to be synthesized. This variable is used throughout the script, thus making the rest of the script generic. By re-defining the value of *active_design* to other sub-modules (*tap_instruction* and *tap_state*), the same script may be used to synthesize the sub-modules. Users can apply the same concept to clock names, clock periods, etc. in order to parameterize the scripts.

Lets assume that you have successfully synthesized three sub-blocks, namely *tap_bypass*, *tap_instruction* and *tap_state*. We can apply the same synthesis script to synthesize the top level, with the exception that we have to include the mapped "db" files for the sub-blocks, before reading the *tap_controller.v* file. Also, the wire-load mode may need to be changed to **enclosed** for proper modeling of the interconnect wires. Since the sub-modules contain the dont_touch attribute, the top-level synthesis will not optimize across boundaries, and may violate the design rule constraints. To remove these violations, you must re-synthesize/optimize the design with the dont_touch attribute removed from the sub-blocks.

DFT scan insertion at the top-level is another reason for removing the dont_touch attribute from the sub-blocks. This is due to the fact that the DFT scan insertion cannot be implemented at the top-level, if the sub-blocks contain the dont_touch attribute. The following script exemplifies this process by performing initial synthesis with scan enabled, before re-compiling (compile –only_design_rule) the design with dont_touch attribute removed from all the sub-blocks.

Synthesis Script for the top-level

```
/***********************************************/
/* Define top-level design name */
    active_design = tap_controller

/***********************************************/
/* Define sub-block design names */
    sub_modules = {tap_bypass tap_instruction tap_state}

    foreach (module, sub_modules) {
        syn_db = module + ".db"
        read syn_db
    }

/***********************************************/
/* Design entry in verilog format */
    analyze –format verilog active_design + ".v"
    elaborate active_design

    current_design active_design
    link

    uniquify
```

```
/************************************************/
/* Design environment */
    set_wire_load LARGE –mode enclosed
    set_operating_conditions WORST

/************************************************/
/* Clock specification and design constraints */
    create_clock –period 33  –waveform {0 16.5} tck
    set_dont_touch_network {tck trst}
    set_clock_skew –delay 2.0 –minus_uncertainty 3.0 tck

    set_driving_cell –cell BUFF1X –pin Z all_inputs()
    set_drive 0 {tck trst}

    set_input_delay   20.0 –clock tck –max all_inputs()
    set_output_delay  10.0 –clock tck –max all_outputs()

    set_max_area 0

/************************************************/
/* Initial compilation with scan insertion */
    set_fix_multiple_port_nets –all –buffer_constants

    compile –scan

    remove_attribute find(–hierarchy design, "*") dont_touch

    current_design active_design
    uniquify

    check_test
    create_test_patterns –sample 10
    preview_scan
    insert_scan
    check_test
```

```
/************************************************/
/* Optional incremental compilation to fix DRC violations */
    compile –only_design_rule

/************************************************/
/* Clean-up and generate the netlist in verilog and db formats */
    remove_unconnected_ports find(–hierarchy cell, "*")
    change_names –hierarchy –rules BORG

    set_dont_touch current_design

    write –hierarchy –output active_design + ".db"
    write –format verilog –hierarchy   \
                        –output active_design + ".sv"
```

2.3.2 Static Timing Analysis using PrimeTime

After successful synthesis, the netlist obtained must be analyzed to check for timing violations. The timing violations may consist of either setup and/or hold-time violations.

The design was synthesized with emphasis on maximizing the setup-time, therefore you may encounter very few setup-time violations, if any. However, the hold-time violations will generally occur at this stage. This is due to the data arriving too fast at the input of sequential cells (data changing its value before being latched by the sequential cells), thereby violating the hold-time requirements.

If the design is failing setup-time requirements, then you have no other option but to re-synthesize the design, targeting the violating path for further optimization. This may involve grouping the violating paths or over-constraining the entire sub-block, which had violations. However, if the design is failing hold-time requirements, you may either fix these violations at the pre-layout level, or may postpone this step until after layout. Many designers prefer the latter approach for minor hold-time violations (also used here), since the pre-layout synthesis and timing analysis uses the statistical wire-load models and fixing the hold-time violations at the pre-layout level may result in setup-time violations for the same path, after layout. It must be noted that gross hold-time violations should be fixed at the pre-layout level, in order to minimize the number of hold-time fixes, which may result after the layout.

PT script for pre-layout setup-time analysis

```
#**************************************************
# Define top-level design name
    set active_design  tap_controller

#**************************************************
# Design entry in db format, netlist only – no constraints
    read_db –netlist_only $active_design.db

    current_design $active_design

#**************************************************
# Design environment
    set_wire_load large
    set_wire_load_mode top

    set_operating_conditions  WORST

    set_load 50.0 [all_outputs]
    set_driving_cell –cell BUFF1X –pin Z [all_inputs]

#**************************************************
# Clock specification and design constraints
    create_clock –period 33 –waveform [0 16.5] tck
    set_clock_latency     2.0 [get_clocks tck]
    set_clock_transition 0.2 [get_clocks tck]
    set_clock_uncertainty 3.0 –setup [get_clocks tck]

    set_input_delay    20.0 –clock tck [all_inputs]
    set_output_delay 10.0 –clock tck [all_outputs]
```

TUTORIAL 25

```
#*************************************************
# Timing analysis commands
    report_constraint –all_violators

    report_timing  –to [all_registers –data_pins]
    report_timing  –to [all_outputs]
```

The above PT script performs the static timing analysis for the *tap_controller* design. Notice that the clock latency and transition are fixed in the above example, because at the pre-layout level the clock tree has not been inserted. Therefore, it is necessary to define a certain amount of delay that approximates the final delay associated with the clock tree. Also, the clock transition is specified because of the high fanout associated with the clock network. The high fanout suggests that the clock network is driving many flip-flops, each having a certain amount of pin capacitance. This gives rise to slow input ramp time for the clock. The fixed transition value (again approximating the final clock tree number) of clock prevents PT from calculating incorrect delay values, that are based upon the slow input ramp to the flops.

The script to perform the hold-time analysis at the pre-layout level is shown below. To check for hold-time violations, the analysis must be performed utilizing the best-case operating conditions, specified in the *ex25_best.db* library. In addition, an extra argument (–delay_type min) is specified in the report_timing command, as follows:

PT script for pre-layout hold-time analysis

```
#*************************************************
# Define top-level design name
    set active_design  tap_controller

#*************************************************
# Design entry in db format, netlist only – no constraints
    read_db –netlist_only $active_design.db

    current_design $active_design
```

```
#***************************************************
# Design environment
    set_wire_load  large
    set_wire_load_mode  top

    set_operating_conditions  BEST

    set_load  50.0  [all_outputs]
    set_driving_cell  –cell  BUFF1X –pin Z  [all_inputs]

#***************************************************
# Clock specification and design constraints
    create_clock  –period  33  –waveform  [0 16.5]  tck
    set_clock_latency     2.0  [get_clocks tck]
    set_clock_transition  0.2  [get_clocks tck]
    set_clock_uncertainty 0.2  –hold  [get_clocks tck]

    set_input_delay   0.0 –clock  tck  [all_inputs]
    set_output_delay  0.0 –clock  tck  [all_outputs]

#***************************************************
# Timing analysis commands
    report_constraint –all_violators

    report_timing  –to [all_registers –data_pins]   \
                                            –delay_type min
    report_timing  –to [all_outputs] –delay_type min
```

2.3.3 SDF Generation

To perform timing simulation, you will need the SDF file for back annotation. The static timing was performed using PT; therefore it is prudent that the SDF file be generated from PT itself. However, most designers feel comfortable in using DC to generate the SDF file. We will therefore use DC to generate the SDF in this section.

Depending on the design, the resultant SDF file may require a certain amount of "massaging" before it can be used to perform timing simulation of the design. The reason for massaging is explained in detail in Chapter 10.

The previous section described a method of defining the clock latency and transition in PT. Similar sets of commands exist for DC. For example:

dc_shell> create_clock –period 33 –waveform {0 16.5} tck

dc_shell> set_clock_skew –delay 2.0 tck

dc_shell> set_clock_transition 0.2 tck

The following script uses the above commands and may be used to generate the pre-layout SDF for the *tap_control*ler design. This SDF file is targeted for simulating the design dynamically with timing. In addition, the script also generates the timing constraints file. Though this file is also in SDF format, it is solely used for forward annotating the timing information to the layout tool in order to perform timing driven layout.

DC script for pre-layout SDF generation

```
active_design = tap_controller

read active_design + ".db"

current_design active_design
link

set_wire_load LARGE –mode top
set_operating_conditions WORST

create_clock –period 33 –waveform {0 16.5} tck
set_clock_skew –delay 2.0 –minus_uncertainty 3.0 tck
set_clock_transition 0.2 tck
```

```
set_driving_cell –cell BUFF1X –pin Z all_inputs()
set_drive 0 {tck trst}

set_load 50 all_outputs()

set_input_delay    20.0 –clock tck –max all_inputs()
set_output_delay   10.0 –clock tck –max all_outputs()

write_timing –format sdf-v2.1       \
             –output active_design + ".sdf"

write_constraints –format sdf –cover_design   \
             –output constraints.sdf
```

2.3.4 Verification

The final step before layout is to verify the structural netlist for functionality. There are two methods that perform this step:

a) Dynamic timing simulation using the SDF file.
b) Formal verification.

2.3.4.1 Gate-Level Simulation with Pre-Layout SDF

This topic needs no clarification, as everyone in the ASIC world is well versed to the process of dynamic simulation. Simulating the structural netlist involves exercising combinations of input signals, while observing the outputs. This is generally performed by means of a test-bench coded in Verilog or VHDL, which emulates the whole system. In other words, it incorporates/envelops the *tap_controller* design.

Dynamic simulation of the gate-level is the traditional approach and is still regarded by many as the best way to debug the design, and to check for functionality at the same time. However, this approach has some shortfalls. Primarily that certain paths of the logic in the design may never be active (get sensitized), for the specified combination of input signals. This means that these paths will not be checked for correct functionality and timing.

Another problem with the dynamic simulation method is the run-time for large designs containing millions of gates. The run-time for these designs can be enormous. It may even become impractical to run gate-level simulations on such designs. The concept of formal verification was introduced to overcome these issues, which are discussed in the next section.

Let us assume that the *tap_controller* design is passing all the functional vectors with no setup or hold-time violations, using the pre-layout SDF generated previously.

2.3.4.2 Formal Verification

Synopsys recently introduced a tool that is capable of performing formal verification, called Formality. Formality uses mathematical algorithms in order to check the logic equivalency of a design. Since this technique is static (i.e., it does not use input test vectors), the verification run-times are reduced dramatically as compared to the dynamic simulation method. Furthermore, it also provides a powerful means to verify large designs in a comparatively short duration. However, the most significant advantage of using Formality is that complete verification of the design is achieved without omitting any logic paths (as is possible by dynamic simulation techniques).

This section (and the book) briefly outlines the purpose and importance of using formal verification techniques and where they can be applied. However, the syntax and usage of the commands used by Formality will not be described.

Formality may be used to verify RTL against RTL, RTL against synthesized gate-level netlist, or gate-level against gate-level netlist. At this point, Formality should be used to verify the RTL against synthesized netlist to check for the functional validity of the gate-level netlist. Compared to gate-level simulation using the pre-layout SDF, Formality takes a fraction of time to completely verify the design.

2.4 Floorplanning and Routing

The floorplanning step involves physical placement of cells and clock tree synthesis. Both these steps are performed within the layout tool. The placement step may include timing driven placement of the cells, which is performed by annotating the *constraints.sdf* file (generated by DC) to the layout tool. This file consists of path delays that include the cell-to-cell timing information. This information is used by the layout tool to place cells with timing as the main criterion i.e., the layout tool will place timing critical cells closer to each other in order to minimize the path delay.

Let's assume that the design has been floorplanned. Also, the clock tree has been inserted in the design by the layout tool. The clock tree insertion modifies the existing structure of the design. In other words, the netlist in the layout tool is different from the original netlist present in DC. This is because of the fact that the design present in the layout tool contains the clock tree, whereas the original design in DC does not contain this information. Therefore, the clock tree information should somehow be transferred to the design residing in DC or PT. The new netlist (containing the clock tree information) should be formally verified against the original netlist to ensure that the transfer of clock tree did not break the functionality of the original logic. Various methods of transferring the clock tree information to the design are explored in detail in Chapter 9. For the sake of simplicity, let us assume that the clock tree information is present in the *tap_controller* design.

The design is now ready for routing. In a broad sense, routing is performed in two phases – global route and detailed route. During global route, the router divides the layout surface into separate regions and performs a point-to-point "loose" routing without actually placing the geometric wires. The final routing is performed by the detailed router, which physically places the geometric wires and routes within the regions. Full explanations of these types of routing are explained in Chapter 9. Lets assume that the design has been global routed.

The next step involves extracting the <u>estimated</u> parasitic capacitances, and RC delays from the global routed design. This step reduces the synthesis-layout iteration time, especially since cell placement and global routing may take much less time than detailed routing the entire chip. However, if the cells are placed optimally with minimal congestion, detailed routing is also very fast. In any case, extraction of delays after the global route phase (albeit estimates) provides a faster method of getting closer to the real delay values that are extracted from the layout database after the detailed routing phase.

Back annotate the estimates to the design in PT for setup and hold-time static timing analysis, using the following scripts.

PT script for setup-time analysis, using estimated delays

```
#************************************************
# Define top-level design name
    set active_design  tap_controller

#************************************************
# Design entry in db format, netlist only – no constraints
    read_db –netlist_only $active_design.db

    current_design  $active_design

#************************************************
# Design environment
    set_operating_conditions  WORST

    set_load  50.0  [all_outputs]
    set_driving_cell  –cell BUFF1X –pin Z  [all_inputs]

#************************************************
# Back annotation of estimated data from floorplan
    source capacitance.pt  # estimated parasitic capacitances
    read_sdf rc_delays.sdf # estimated RC delays

#************************************************
# Clock specification and design constraints
    create_clock  –period 33  –waveform  [0 16.5]  tck
    set_propagated_clock [get_clocks tck]
    set_clock_uncertainty  0.5  –setup [get_clocks tck]
```

```
set_input_delay   20.0 –clock tck [all_inputs]
set_output_delay  10.0 –clock tck [all_outputs]
```

#**
Timing analysis commands
```
report_constraint –all_violators

report_timing –to [all_registers –data_pins]
report_timing –to [all_outputs]
```

PT script for hold-time analysis, using estimated delays

#**
Define design name
```
set active_design  tap_controller
```

#**
Design entry in db format, netlist only – no constraints
```
read_db –netlist_only $active_design.db

current_design $active_design
```

#**
Design environment
```
set_operating_conditions  BEST

set_load 20.0 [all_outputs]
set_driving_cell –cell BUFF1X –pin Z [all_inputs]
```

#**
Back annotation of estimated data from floorplan
```
source capacitance.pt   # estimated parasitic capacitances
read_sdf rc_delays.sdf  # estimated RC delays
```

```
#***************************************************
# Clock specification and design constraints
    create_clock –period 33 –waveform [0 16.5] tck
    set_propagated_clock [get_clocks tck]
    set_clock_uncertainty 0.05 –hold [get_clocks tck]

    set_input_delay   0.0 –clock tck [all_inputs]
    set_output_delay  0.0 –clock tck [all_outputs]

#***************************************************
# Timing analysis commands
    report_constraint –all_violators

    report_timing  –to [all_registers –data_pins]   \
                                        –delay_type min
    report_timing  –to [all_outputs] –delay_type min
```

The above script back annotates *capacitance.pt* and *rc_delays.sdf* file. The *capacitance.pt* file contains the capacitive loading per net of the design in set_load format, while the *rc_delays.sdf* file contains point-to-point interconnect RC delays of individual nets. DC (and PT) performs the calculation of cell delay, based upon the output net loading and input slope of each cell in the design. The reason for using this approach is explained in detail in Chapter 9.

If the design fails setup-time requirements, you may re-synthesize the design with adjusted constraints or re-floorplan the design. If the design is failing hold-time requirements, then depending on the degree of violation you may decide to proceed to the final step of detailed routing the design, or re-optimize the design with adjusted constraints.

If re-synthesis is desired then the floorplan (placement) information consisting of the physical clusters and cell locations, should be back annotated to DC. This step is desired because up till now, DC did not know the physical placement information of cells. By annotating the placement information to DC, the post-layout optimization of the design within DC is vastly improved. The layout tool generates the physical information in PDEF format that can be read by DC, using the following command:

> read_clusters <file name in PDEF format>

The script to perform this is similar to the initial synthesis script, with the exception of the annotated data and incremental optimization of the design, as illustrated below:

Script for incremental synthesis of the design

```
active_design = tap_controller

read  active_design + ".db"

current_design active_design
link

include capacitance.dc   /* estimated parasitic capacitances */
read_timing –f sdf rc_delays.sdf   /* estimated RC delays */
read_clusters clusters.pdef /* physical information */

/*****************************************************************/
/* create custom wire-load models for next iteration of synthesis */
    create_wire_load –hierarchy           \
                    –percentile 80        \
                    –output cwlm.txt

create_clock –period 33 –waveform {0 16.5} tck
set_clock_skew  –propagated –minus_uncertainty 3.0  tck
set_dont_touch_network {tck  trst}

set_driving_cell –cell BUFF1X –pin Z all_inputs()
set_drive 0 {tck trst}

set_input_delay    20.0 –clock  tck –max  all_inputs()
set_output_delay  10.0 –clock  tck –max all_outputs()
```

```
            set_max_area 0

            set_fix_multiple_port_nets –all –buffer_constants

/**********************************************************************/
/* Re-optimize the design to control the amount of modifications */
            reoptimize_design –in_place

            write –hierarchy –output active_design + ".db"
            write –format verilog –hierarchy \
                            –output active_design + ".sv"
```

The create_wire_load command used in the above script creates a custom wire-load model for the *tap_controller* design. The initial synthesis run used the wire-load models present the in the technology library that are not design specific. Therefore, in order to achieve better accuracy for the next synthesis iteration, the custom wire-load models specific to the design should be used.

The following command may be used to update the technology library present in DC's memory to reflect the new custom wire-load models. For example:

dc_shell> update_lib ex25_worst.db cwlm.txt

Let's assume that the design has been re-analyzed and is now passing both setup and hold-time requirements. The next step is to detail route the design. This is a layout dependent feature, therefore will not be discussed here.

2.5 Post-Layout Steps

The post-layout steps involve, verifying the design for timing with actual delays back annotated; functional simulation of the design; and lastly, performing LVS and DRC.

Let us presume that the design has been fully routed with minimal congestion and area. The finished layout surface must then be extracted to get the actual parasitic capacitances and interconnect RC delays. Depending upon the layout tool and the type of extraction, the extracted values are generally written out in the SDF format for the interconnect RC delays, while the parasitic information is generated as a string of set_load commands for each net in the design. In addition, if a hierarchical place and route has been performed, the physical placement location of cells in the PDEF format should also be generated.

2.5.1 Static Timing Analysis using PrimeTime

The first step after layout is to perform static timing on the design, using the actual delays. Similar to post-placement, the post-route timing analysis uses the same commands, except that this time the <u>actual</u> delays are back annotated to the design.

Predominantly, the timing of the design is dependent upon clock latency and skew. It is therefore prudent to perform the clock skew analysis before attempting to analyze the whole design. A useful Tcl script is provided by Synopsys through their on-line support on the web, called SolvNET. You may download this script and run the analysis before proceeding. Let us assume that the clock latency and skew is within limits. The next step is to perform the static timing analysis on the design, to check the setup and hold-time violations (if any) using the following scripts:

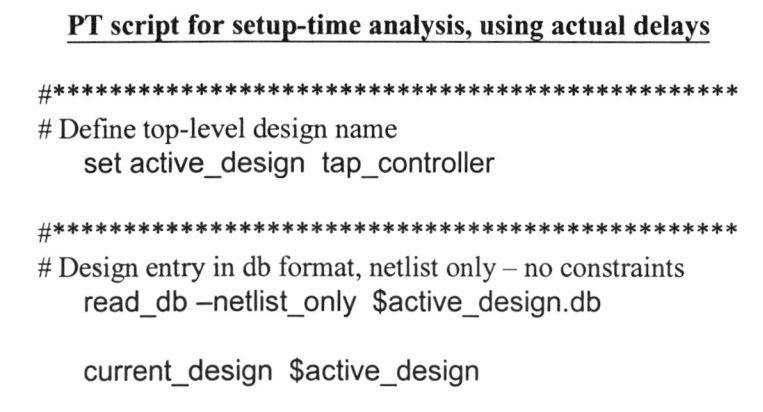

PT script for setup-time analysis, using actual delays

```
#************************************************
# Define top-level design name
    set active_design  tap_controller

#************************************************
# Design entry in db format, netlist only – no constraints
    read_db –netlist_only $active_design.db

    current_design $active_design
```

```
#************************************************
# Design environment
    set_operating_conditions  WORST

    set_load  50.0 [all_outputs]
    set_driving_cell  –cell  BUFF1X –pin Z [all_inputs]

#************************************************
# Back annotation of real data from the routed design
    source capacitance.pt   # actual parasitic capacitances
    read_sdf rc_delays.sdf  # actual RC delays
    read_parasitics clock_info_wrst.spf # for clocks etc.

#************************************************
# Clock specification and design constraints
    create_clock  –period  33  –waveform  [0 16.5]  tck
    set_propagated_clock  [get_clocks tck]
    set_clock_uncertainty  0.5  –setup [get_clocks tck]

    set_input_delay    20.0 –clock  tck  [all_inputs]
    set_output_delay  10.0 –clock  tck  [all_outputs]

#************************************************
# Timing analysis commands
    report_constraint –all_violators

    report_timing  –to [all_registers –data_pins]
    report_timing  –to [all_outputs]
```

PT script for hold-time analysis, using actual delays

```
#************************************************
# Define top-level design name
    set active_design  tap_controller
```

```
#***************************************************
# Design entry in db format, netlist only – no constraints
    read_db –netlist_only $active_design.db

    current_design $active_design

#***************************************************
# Design environment
    set_operating_conditions  BEST

#***************************************************
# Back annotation of real data from the routed design
    source capacitance.pt  # actual parasitic capacitances
    read_sdf rc_delays.sdf # actual RC delays
    read_parasitics clock_info_best.spf # for clocks etc.

    set_load 50.0 [all_outputs]
    set_driving_cell –cell BUFF1X –pin Z [all_inputs]

#***************************************************
# Clock specification and design constraints
    create_clock –period 33 –waveform [0 16.5] tck
    set_propagated_clock [get_clocks tck]
    set_clock_uncertainty 0.05 –hold [get_clocks tck]

    set_input_delay   0.0 –clock tck [all_inputs]
    set_output_delay  0.0 –clock tck [all_outputs]

#***************************************************
# Timing analysis commands
    report_constraint –all_violators

    report_timing  –to [all_registers –data_pins]   \
                                         –delay_type min
    report_timing  –to [all_outputs] –delay_type min
```

2.5.2 Post-Layout Optimization

The post-layout optimization or PLO may be performed on the design to improve or fix the timing requirements. DC provides several methods of fixing timing violations, through the in-place optimization (or IPO) feature. As before, DC also makes use of the physical placement information to perform location based optimization (LBO). In this example, we will use the cell resizing and buffer insertion feature of the IPO to fix the hold-time violations.

2.5.2.1 Fixing Hold-Time Violations

The design was synthesized for maximum setup-time requirements. Timing was verified at each step (after synthesis and then, after the global route phase), therefore in all probability the routed design will pass the setup-time requirements. However, some parts of the design may fail hold-time requirements at various endpoints.

If the design fails the hold-time requirements then you should fix the violations by adding buffers to delay the arrival time of the failing signals, with respect to the clock. Let's assume that the design is failing hold-time requirements at multiple endpoints.

There are various approaches to fix the hold-time violations. Such methods are discussed in detail in Chapter 9. In this example, we will utilize the dc_shell commands to fix the hold time violations, as illustrated below:

DC Script to fix the hold-time violations

```
active_design = tap_controller

read active_design + ".db"

current_design active_design
link
```

```
include capacitance.dc   /*actual parasitic capacitances */
read_timing –f sdf rc_delays.sdf    /*actual RC delays */
read_clusters clusters.pdef   /*physical hierarchy info */

create_clock –period 33 –waveform {0 16.5} tck
set_clock_skew –propagated –plus_uncertainty 0.05 tck

set_dont_touch_network {tck trst}

set_driving_cell –cell BUFF1X –pin Z all_inputs()
set_drive 0 {tck trst}

set_input_delay    –min 0.0 –clock tck –max all_inputs()
set_output_delay   –min 0.0 –clock tck –max all_outputs()

set_fix_hold tck  /* fix hold-time violations w.r.t. tck */

reoptimize_design –in_place

write –hierarchy –output active_design + ".db"
write –format verilog –hierarchy \
                 –output active_design + ".sv"
```

In the above script, the set_fix_hold command instructs DC to fix hold-time violations with respect to the clock *tck*. The –in_place argument of the reoptimize_design command is the IPO command, which is regulated by various variables that are described in Chapter 9. Making use of these variables, DC inserts or resizes the gates to fix the hold time violations. The LBO variables are helpful in inserting the buffers at the correct location, so as to minimize its impact on some other logic path, leading off from the violating path.

After IPO, the design should again be analyzed through PT to ensure that the violations have been fixed using the post-layout PT script illustrated before.

TUTORIAL

Once the design passes all timing requirements, the post-layout SDF may be generated (from PT or DC) for simulation purposes, if needed. We will use DC to generate the worst-case post-layout SDF using the script provided below. A similar script may be used to generate the best-case SDF. Obviously, you need to back annotate the best-case extracted numbers from the layout tool to generate the best-case SDF from DC. This solely depends on the layout tool and the methodology being used.

DC script for worst-case post-layout SDF generation

```
active_design = tap_controller

read active_design + ".db"

current_design active_design
link

set_operating_conditions WORST

include capacitance.dc   /* actual parasitic capacitances */
read_timing rc_delays.sdf   /* actual RC delays */

create_clock –period 33 –waveform {0 16.5} tck
set_clock_skew –propagated –minus_uncertainty 0.5  tck

set_driving_cell –cell BUFF1X –pin Z all_inputs()
set_drive 0 {tck trst}

set_load 50 all_outputs()

set_input_delay    20.0 –clock  tck –max  all_inputs()
set_output_delay  10.0 –clock  tck –max all_outputs()

write_timing –format sdf-v2.1    \
             –output active_design + ".sdf"
```

It is recommended that formal verification be performed again between the source RTL and the final netlist, to check for any errors that may have been unintentionally introduced during the whole process. This is the final step; the design is now ready for LVS and DRC checks, before tape-out.

2.6 Chapter Summary

This chapter highlighted the practical side of the ASIC design methodology in the form of a tutorial. An example design was used to guide the reader from start to finish. At each stage, brief explanation and relevant scripts were provided.

The chapter started with basics of setting up the Synopsys environment and technical specification of the example design. Further sections were divided into pre-layout, floorplanning and routing, and finally the post-layout steps.

The pre-layout steps included initial synthesis and scan insertion of the design, along with static timing analysis, and SDF generation for dynamic simulation. In order to minimize the synthesis-layout iterations, the floorplanning and routing section stressed upon the placement of cells, with emphasis on back annotating to DC, the estimated delays extracted after global routing the design. The final section used post-layout optimization techniques to fix the hold-time violations, and to generate the final SDF for simulation.

Finally, the application of formal verification method using Synopsys Formality was also included. This section did not contain any scripts, but the reader was made aware of the usefulness of formal techniques and where they are applied.

3

BASIC CONCEPTS

This chapter covers the basic concepts related to synthesis, using Synopsys suite of tools. These concepts introduce the reader to synthesis terminology used throughout the later chapters. These terms provide the necessary framework for Synopsys synthesis and static timing analysis.

Although this chapter is a good reference, advanced users of Synopsys tools, already familiar with Synopsys terminology, may skip this chapter.

3.1 Synopsys Products

This section briefly describes all relevant Synopsys products related to this book.

a) Library Compiler
b) Design Compiler and Design Analyzer
c) PrimeTime
d) Test Compiler
e) Formality

Library Compiler
The core of any ASIC design is the technology library containing a set of logic cells. The library may contain functional description, timing, area and other pertinent information of each cell. Library Compiler (LC) parses this textual information for completeness and correctness, before converting it to a format, used globally by all Synopsys applications.

Library Compiler is invoked by typing lc_shell in a UNIX shell. All the capabilities of the LC can also be utilized within dc_shell.

Design Compiler and Design Analyzer
The Synopsys Design Compiler (DC) and Design Analyzer (DA) comprise a powerful suite of logic synthesis products, designed to provide an optimal gate-level synthesized netlist based on the design specifications, and timing constraints. In addition to high-level synthesis capabilities, it also incorporates a static timing analysis engine, along with solutions for FPGA synthesis and links-to-layout (LTL).

Design Compiler is the command line interface of Synopsys synthesis tool and is invoked by typing dc_shell in a UNIX shell. The Design Analyzer is the graphical front-end version of DC and is launched by typing design_analyzer. Design Analyzer also supports schematic generation, with critical path analysis through point-to-point highlighting.

Although, beginners may initially prefer using DA, they quickly migrate to using DC, as they become more familiar with Synopsys commands.

PrimeTime
PrimeTime (PT) is the Synopsys sign-off quality, full chip, gate-level static timing analysis tool. In addition, it also allows for comprehensive modeling capabilities, often required by large designs.

PT is faster compared to DC's internal static timing analysis engine. It also provides enhanced analysis capabilities, both textually and graphically. In contrast to the rest of Synopsys tools, this tool is Tcl language based, therefore providing powerful features of that language to promote the analysis and debugging of the design.

PT is a stand-alone tool and can be invoked as a command line interface or graphically. To use the command line interface, type **pt_shell** in the UNIX window, or type **primetime** for the graphical version.

Test Compiler
The Test Compiler (TC) is the Synopsys test insertion tool that is incorporated within the DC suite of tools. The TC is used to insert DFT features like scan insertion and boundary scan, to the design. All TC commands are directly invoked from **dc_shell**.

Formality
Formality is the Synopsys formal verification tool. This tool was introduced recently by Synopsys. It is fully compatible with all Synopsys formats. The tool features enhanced graphical debugging capabilities that include schematic representation of logic under verification, and visual suggestions annotated to the schematic as pointers of possible incorrect logic. It also provides suggestions for possible fixes to the design.

3.2 Synthesis Environment

As with most EDA products, Synopsys tools require a setup file that specifies the technology library location and other parameters used for synthesis. Synopsys also defines its own format for storing and processing the information. This section highlights such details.

3.2.1 Startup Files

There is a common startup file called ".synopsys_dc.setup" for all tools of the DC family. A separate startup file is required for PT which is named ".synopsys_pt.setup". This file is in Tcl format, and contains path information to the technology libraries and parameters required by PT.

The default startup files for both DC and PT reside in the Synopsys installation directory, and are automatically loaded upon invocation of these tools. These default files do not contain the design dependent data. Their function is to load the Synopsys technology independent libraries and other parameters. The user in the startup files specifies the design dependent data. During startup, DC or PT reads these files in the following order:

1. Synopsys installation directory.
2. Users home directory.
3. Project working directory.

The settings specified in the startup files, residing in the project working directory, override the ones specified in the home directory and so forth, i.e., the configuration specified in the project working directory takes precedence over all other settings.

It is up to the discretion of the user to keep these files wherever it is convenient. However, it is recommended that the design dependent startup files be kept in the working directory.

The minimum information required by Design Compiler, is the search_path, target_library, link_library and the symbol_library. PT requires the search_path and link_path information only. Typical startup files are shown in Example 3.1. Note the change in format between the DC and PT setup files.

Example 3.1   Synopsys Setup Files

DC .synopsys_dc.setup file

```
company     = "Lucknow Technologies";
designer    = "Clemson";
technology  = "0.25 micron";

search_path = search_path + {   "."    "/usr/golden/library/std_cells" \
                                       "/ usr/golden/library/pads"}

target_library  = {std_cells_lib.db}
link_library    = {"*", std_cells_lib.db, pad_lib.db}
symbol_library  = {std_cells_lib.sdb, pad_lib.sdb}
```

PT .synopsys_pt.setup file

```
set search_path    [list . /usr/golden/library/std_cells    \
                          /usr/golden/library/pads]
set link_path      [list {*} std_cells_lib.db pad_lib.db]
```

3.2.2 System Library Variables

At this time, it is worth explaining the difference between the target_library and the link_library system variables. The target_library specifies the name of the technology library that corresponds to the library whose cells the designers want DC to infer and finally map to. The link_library defines the name of the library that refers to the library of cells used solely for reference, i.e., cells in the link_library are not inferred by DC. For example, you may specify a standard cell technology library as the target_library, while specifying the pad technology library name and all other macros (RAMs, ROMs etc.) in the link_library list. This means that the user would synthesize the design that targets the cells present in the standard cell library, while linking to the pads and macros that are instantiated in the design. If the pad library is included in the target_library list, then DC may use the pads to synthesize the core logic.

The target library name should also be included in the link_library list, as shown in Example 3.1. This is important while reading the gate-level netlist in DC. DC will not be able to link to the mapped cells in the netlist, if the target library name is not included in the link library list. For this case, DC generates a warning stating that it was unable to resolve reference for the cells present in the netlist.

The target_library and link_library system variables allow the designer to better control the mapping of cells. These variables also provide a useful means to re-map a gate-level netlist from one technology to the other. In this case, the link_library may contain the old technology library name, while the target_library may contain the new technology library. Re-mapping can be performed by using the translate command in dc_shell.

The symbol_library system variable holds the name of the library, containing graphical representation of the cells in the technology library. It is used, to represent the gates schematically, while using the graphical front-end tool, DA. The symbol libraries are identified with a "sdb" extension. If this variable is omitted from the setup file, DA will use a generic symbol library called "generic.sdb" to create schematics. Generally, all technology libraries provided by the library vendor include a corresponding symbol library. It is imperative that there be a exact match of the cell names and the pin names, between the technology and the symbol library. Any mismatch in a cell will cause DA to reject the cell from the symbol library, and use the cell from the generic library.

It must be noted that DC uses the link_library variable, whereas PT calls it the link_path. Apart from the difference in name and the format, the application of both these variables is identical. Since PT is a gate-level static timing analyzer, it only works on the structural gate-level netlists. Thus, PT does not utilize the target_library variable.

3.3 Objects, Variables and Attributes

Synopsys supports a number of objects, variables and attributes in order to streamline the synthesis process. Using these, designers can write powerful dc_shell scripts to automate the synthesis process. It is therefore essential for designers to familiarize themselves with these terms.

3.3.1 Design Objects

There are eight different types of design objects categorized by DC. These are:

- *Design*: It corresponds to the circuit description that performs some logical function. The design may be stand-alone or may include other sub-designs. Although, sub-designs may be part of the design, it is treated as another design by Synopsys.

BASIC CONCEPTS

- *Cell*: It is the instantiated name of the sub-design in the design. In Synopsys terminology, there is no differentiation between the cell and instance; both are treated as cell.

- *Reference*: This is the definition of the original design to which the cell or instance refers. For e.g., a leaf cell in the netlist must be referenced from the link library, which contains the functional description of the cell. Similarly, an instantiated sub-design (called cell by Synopsys) must be referenced in the design, which contains functional description of the instantiated sub-design.

- *Port*: These are the primary inputs, outputs or IO's of the design.

- *Pin*: It corresponds to the inputs, outputs or IO's of the cells in the design (Note the difference between port and pin)

- *Net*: These are the signal names, i.e., the wires that hook up the design together by connecting ports to pins and/or pins to each other.

- *Clock*: The port or pin that is identified as a clock source. The identification may be internal to the library or it may be done using dc_shell commands.

- *Library*: Corresponds to the collection of technology specific cells that the design is targeting for synthesis; or linking for reference.

3.3.2 Variables

Variables are placeholders used by DC for the purpose of storing information. The information may relate to instructions for tailoring the final netlist, or it may contain user-defined value to be used for automating the synthesis process. Some variables are pre-defined by DC and may be used by the designer to obtain the current value stored in the variable. For e.g., the variable called "dc_shell_status" has a special meaning to DC, while "captain_picard" has no meaning to DC. The latter may be used to hold any user-defined value for scripting purposes.

All variables are global and last only during the session. They are not saved along with the design database. Upon completion of the dc_shell session, the value of the variables is lost. Most dc_shell variables have a default value associated with them, which they inherit at the start of the session. For instance, the following variable uses IEEE.std_logic_1164 as its default value and you need to specify additional values (packages in this case) for DC to write out the "USE" clause with the specified name in the synthesized VHDL netlist. For example:

dc_shell> vhdlout_use_packages = { library IEEE.std_logic_1164; \
library STD_LIB; }

Users may use variables to gather useful information about the design. Example 3.2 provides a dc_shell script, aliased as *reg_count*, which may be used to find the number of flops per clock domain. To use the following script, the clock name in the design should first be assigned to the variable named CLK.

Example 3.2

```
alias reg_count "list CLK > /dev/null;                              \
    if (dc_shell_status !=1)                                        \
        {echo "Set the CLK variable for a chosen clock";}           \
    else {create_clock CLK –name CLK –period 100;                   \
        remove_variable count > /dev/null;                          \
        count = 0 > /dev/null;                                      \
            foreach(reg, all_registers (-clock CLK)) {              \
            count = count + 1;                                      \
            };                                                      \
    echo "Number of flops for the chosen clock is: " count;         \
        }"
```

A list of all DC variables may be obtained by using the following dc_shell command:

list –variables all

Variables may be removed from DC by using the following dc_shell command:

> remove_variable <variable name>

The variables used by PT differ from those of DC, in syntax and application. PT uses the Tcl based syntax, while DC uses its own. The applications of the PT variables are based on Tcl constructs. Usage of PT variables through Tcl is described in detail in Chapter 11.

3.3.3 Attributes

Attributes are similar in nature to variables. Both store information. However, attributes store information on a particular design object such as nets, cells or clocks.

Generally, attributes are pre-defined and have special meaning to DC, though designers may set their own attributes if desired. For e.g., the set_dont_touch is a pre-defined attribute, used to set a dont_touch on a design, thereby disabling DC optimization on that design.

Attributes are set on and retrieved from the design object by using the following commands:

> set_attribute <object list>
>   <attribute name>
>   <attribute value>
>
> get_attribute <object list>
>   <attribute name>

For example, one may use the following command to find the max_transition value set in the library (called "STD_LIB") and the area attribute set for an inverter (INV2X) in that library:

dc_shell> get_attribute STD_LIB default_max_transition

```
dc_shell> get_attribute STD_LIB/INV2X   area
```

The "**help attributes**" command in dc_shell provides a complete listing of attributes used by DC.

3.4 Finding Design Objects

One of the most useful commands provided by DC is the find command. Sometimes, it becomes necessary to locate objects in dc_shell for the purpose of scripting or automating the synthesis process. The find command is used to locate a list of designs or library objects in DC.

>   find <type> <name list> –hierarchy

The *type* field refers to the object type: design, port, reference, cell, clock, net, pin or library. The *name list* is an optional argument and as the term "name" suggests, is used to specify the design or library object present in dc_shell. If omitted, all objects of the specified type are returned. One of the most important arguments of the find command is the –hierarchy option. This is used to force the find command to traverse the whole hierarchy in search of the specified object.

This command is also useful in explicitly identifying the object type (net, port etc.) and name, to which a particular constraint or attribute is being applied. For instance, consider a design that contains a port and a net having the same name, say "enterprise". Annotating a value, using the set_load command on "enterprise" causes a problem, because the name "enterprise" does not specify a unique object. To resolve this conflict, the user may use the find command in conjunction with the set_load command, to explicitly identify the object type the set_load is be applied to, as follows:

```
dc_shell> set_load 2.0   find(net, "enterprise")
```

The usefulness of the find command is highlighted by its ability to accept wildcard characters, in case the user does not know the full name of the object. This feature makes this command extremely powerful, when used for

BASIC CONCEPTS   53

scripting and debugging purposes. For example, one may use this command to apply a set_dont_touch attribute on all sub-designs whose name begin with subA_ and subB_ , as follows:

`dc_shell>` set_dont_touch find(design, {subA_* subB_*})

The following examples further illustrate the usage of this command:

`dc_shell>` find port

☺ Lists all the ports of the current_design.

`dc_shell>` set_dont_touch_network find(net, {"gen_clk", "scan"})

☺ Applies the dont_touch_network attribute on the specified nets.

`dc_shell>` remove_attribute find(design, "*" –hierarchy) dont_touch

☺ Removes the dont_touch attribute from the whole design throughout the hierarchy

`dc_shell>` find (pin, stdcells_lib/DFF1/*)

☺ Lists all the pins of DFF1 cell present in the library called *stdcells_lib*.

Similar to DC, PT also provides this capability, albeit differently. PT uses the get_* commands to find objects in a design. The objects to be found are encompassed within the get command. For example:

get_ports, get_nets, get_designs, get_lib_cells, get_cells etc.

A full list of get commands is listed in the PrimeTime User Guide.

3.5 Synopsys Formats

Most Synopsys products support and share, a common internal structure, called the "db" format. The db files are the binary compiled forms representing the text data, be it the RTL code, the mapped gate-level designs, or the Synopsys library itself. The db files may also contain any constraints that have been applied to the design.

In addition, all Synopsys tools understand the following formats of HDL. DC is capable of reading or writing any of these formats.

1. Verilog
2. VHDL
3. EDIF

Today, Verilog and VHDL are the two main HDLs in use, for coding a design. EDIF (Electronic Design Interchange Format) is primarily utilized for porting the gate level netlist, from one tool to another. EDIF was a popular choice a few years back. However, recently Verilog has gained popularity and dominance prompted by its simple to read format and description. Most of the EDA tools today, support both Verilog and EDIF.

VHDL in general is not used for porting the netlist from one vendor tool to another, since it requires the use of IEEE packages, which may vary between different tools. This language is essentially used for the purpose of coding the design and system level verification.

3.6 Data Organization

It is a good practice to organize files according to their formats. This facilitates automating the synthesis process. A common practice is to organize them using the following file extensions.

Script files:	<filename>.scr
RTL Verilog file:	<filename>.v
Synthesized Verilog netlist:	<filename>.sv
RTL VHDL file:	<filename>.vhd
Synthesized VHDL netlist:	<filename>.svhd
EDIF file:	<filename>.edf
Synopsys database file:	<filename>.db
Reports:	<filename>.rpt
Log files:	<filename>.log

3.7 Design Entry

Before synthesis, the design must be entered in DC in the RTL format (although other formats also exist). DC provides the following two methods of design entry:

a) **read** command
b) **analyze & elaborate** commands

Synopsys initially introduced the **read** command, which was then followed by the **analyze/elaborate** commands. The latter commands for design entry provide a fast and powerful method over the **read** command and are recommended for RTL design entry.

The **analyze** and **elaborate** commands are two different commands, allowing designers to initially analyze the design for syntax errors and RTL translation before building the generic logic for the design. The generic logic or GTECH components are part of the Synopsys generic technology independent library. They are unmapped representations of boolean functions and serve as place holders for the technology dependent library.

The **analyze** command also stores the result of the translation in the specified design library (UNIX directory) that may be used later. For example, a design analyzed previously may not need re-analysis and can merely be elaborated, thus saving time. Conversely, the **read** command performs the function of both **analyze** and **elaborate** commands but does not store the analyzed results, therefore making the process slow in comparison.

Parameterized designs (such as usage of *generic* statement in VHDL) must use **analyze** and **elaborate** commands in order to pass required parameters, while elaborating the design. The **read** command should be used for entering pre-compiled designs or netlists in DC.

The following table lists major differences between the **read** and **analyze/elaborate** commands for various categories:

Table 3-1. Difference between analyze/elaborate and read commands

Category	analyze/elaborate	read
Input format	RTL in Verilog or VHDL.	All formats: Verilog, VHDL, EDIF, db etc.
Recommended usage	Synthesizing the RTL in Verilog or VHDL format.	Reading netlists, pre-compiled designs etc.
Design Libraries	Use –library option to specify design library other than the directory from which dc_shell was invoked.	Storing of analyzed results is not possible.
Generics (used in VHDL)	Parameters of the generic statements may be set during elaboration of the design.	Cannot be used to pass parameters.
Architecture (in VHDL)	Can specify architecture in VHDL to be elaborated.	Cannot specify architecture in VHDL to be elaborated.

In contrast to DC, PT uses different commands for design entry. PT, being a static timing analyzer, only works on the mapped structural netlists. The design entry commands used by PT are described in Chapter 11.

3.8 Compiler Directives

Sometimes it is necessary to control the synthesis process from the HDL source itself. This control is primarily needed because of differences that may exist between the synthesis and the simulation environments. Other times, the control is needed simply to direct DC to map to certain types of components; or for embedding the constraints and attributes directly in the HDL source code.

DC provides a number of compiler directives targeted specifically for Verilog and VHDL design entry formats. These directives provide the means to control the outcome of synthesis, directly from the HDL source code. The directives are specified as "comments" in the HDL code, but have specific meaning for DC. These special comments alter the synthesis process, but have no effect on the simulation.

The following sub-sections describe some of the most commonly used directives, both for Verilog and VHDL formats. For a complete list of directives, users are advised to refer to the Design Compiler Reference Manual.

3.8.1 HDL Compiler Directives

The HDL compiler directives refer to the translation process of RTL in Verilog format to the internal format used by Design Compiler. As stated above, specific aspects of the translation are controlled by "comments" within the Verilog source code. At the beginning of each directive is the regular Verilog comment // or /* followed by the keyword "**synopsys**" (all in lower case). Generally, users prefer the former style to specify HDL compiler directives. Therefore, to keep it simple, only // style of comments for HDL directives are discussed in this section.

All comments beginning with // **synopsys** are assumed to only contain HDL compiler directives. DC displays an error if anything apart from HDL compiler directives (say other comments or parts of Verilog code) are present after the // **synopsys** statement.

3.8.1.1 translate_off and translate_on Directives

These are some of the most useful and frequently used directives. They provide the means to instruct DC to stop translation of the Verilog source code from the start of "// **synopsys translate_off**", and start the translation again after it reaches the next directive, "// **synopsys translate_on**". These directives must be used in pairs, with the **translate_off** directive taking the lead.

Consider a scenario where parts of the code present in the source RTL is meant solely for the purpose of dynamic simulation (or maybe the test bench is structured to make use of these statements). Example 3.3 illustrates such a scenario, which contains the Verilog `ifdef` statement to facilitate setting parameters at the command line during simulation. Such a code is clearly unsynthesizable, since the VENDOR_ID depends on the mode specified during simulation. Furthermore, since the HDL compiler cannot handle this statement, it issues an error stating that the design could not be read due to "Undefined macro 'ifdef ....."

Example 3.3

```
`ifdef MY_COMPANY
   `define VENDOR_ID 16'h0083
`else
   `define VENDOR_ID 16'h0036
`endif
```

The translate_off and translate_on HDL directives may be used in this case to bypass the "simulation only" parts of the verilog code as illustrated in Example 3.4. The resulting logic will contain the VENDOR_ID values pertaining to MY_COMPANY only. To change it to the other value, the user has to edit the code and move the HDL directives to make the other VENDOR_ID value visible.

Example 3.4

```
// synopsys translate_off
  `ifdef MY_COMPANY
// synopsys translate_on

     `define VENDOR_ID 16'h0083

// synopsys translate_off
  `else
     `define VENDOR_ID 16'h0036
  `endif
// synopsys translate_on
```

3.8.1.2 parallel_case and full_case Directives

Unlike VHDL – where all conditions within the *case* statement are required to be mutually exclusive – the use of *case* statement in Verilog provides different choices, and may be used to infer a priority encoder. In some cases, the priority encoder is not desired, but one may still want to use the *case* statement to infer a multiplexer. The // synopsys parallel_case directive, forces DC to generate the multiplexer logic, instead of building the priority encoder logic. This directive is generally used for coding a state machine, driven by the fact that a state machine cannot exist in more than one state at a given time.

This directive forces DC to evaluate all *case* conditions in parallel, with execution for all conditions evaluating to true at once. If care is not exercised, this may result in a mismatch between simulation and synthesized logic. To avoid this problem and to make the RTL code more robust, it is suggested that this directive be used when the user knows that only one condition will be evaluated to true at a given time.

The parallel_case directive should be used immediately following the case statement as shown in example 3.5.

Example 3.5

```
always @ (cs_state)
   begin
      case (cs_state)   // synopsys parallel_case
          2'b00:    ns = 2'b01;
          2'b01:    ns = 2'b10;
          2'b10:    ns = 2'b00;
          default: ns = 2'b00;
      endcase
   end
```

The full_case directive is also specific to Verilog HDL. VHDL requires that all conditions in the *case* statement be covered. In contrast, Verilog – because of its "openness" – does not require all conditions of the *case* statement to be covered. Although, Verilog does provide, a *default* clause that is used to cover the rest of the conditions, designers sometimes forget to include this (which is totally acceptable by the Verilog compiler, since it is not a requirement). This causes DC to infer unwanted latches, due to non-assignment of values to the register or net, under all possible conditions. To prevent this, the // synopsys full_case directive should be used, which informs DC, that all conditions are covered.

The // synopsys full_case directive is also used just after the *case* statement, as shown in Example 3.6. Notice that the *default* clause is omitted within the *case* statement.

Example 3.6

```
always @ (SEL or A1 or A2)
   begin
      case (SEL)    // synopsys full_case
           2'b00:     Z = A1;
           2'b01:     Z = A2;
           2'b10:     Z = A1 & A2;
      endcase
end
```

3.8.2 VHDL Compiler Directives

Similar to the HDL compiler, the VHDL compiler directives are special VHDL comments that affect the actions of the VHDL compiler. All VHDL compiler directives start with the VHDL comment (--), followed either by synopsys or pragma statements. This provides a special meaning to the compiler and compels it to perform specified task.

3.8.2.1 translate_off and translate_on Directives

These directives work in the similar fashion as the ones described previously for the HDL compiler, with the exception that these require the VHDL comments as follows:

```
-- synopsys translate_off
-- synopsys translate_on
-- pragma translate_off
-- pragma translate_on
```

The VHDL compiler ignores any RTL code between the translate_off/on directives, however it does perform a syntax check on the embedded code. In order to refrain the compiler from conducting syntax checks; the code must be made completely transparent. This can be achieved by setting the following variable to true:

$$hdlin\_translate\_off\_skip\_text = true$$

These directives are primarily used to block simulation specific constructs in the VHDL code. For example, the user may have a library statement present in the netlist, which specifies the name of the library that contains the VITAL models of the gates present in the netlist. This means that for the purpose of simulation, the gates present in the netlist are being referenced from this library. Upon reading the VHDL code, DC produces an error, since the library statement is specific to simulation only. To circumvent this problem, one may envelop the library statement with the above directives to force DC to completely ignore the library statement.

3.8.2.2 synthesis_off and synthesis_on Directives

The synthesis_off/on directives work in a manner similar to the translate_on/off directives. The behavior of synthesis_off/on directives itself, is not affected by the value of the hdlin_translate_off_skip_text variable. However, the translate_off/on directives perform exactly the same function if the value of the variable specified above is set to false.

The above directives are the preferred approach to hide the simulation only constructs. Though, the VHDL compiler performs the syntax checks of the code present within these directives, it ignores the code for the purpose of synthesis.

Syntactically, these variables may be used as follows:

```
-- pragma synthesis_off
   <VHDL code goes here, used only for simulation>
-- pragma synthesis_on
```

3.9 Chapter Summary

This chapter introduced the reader to various terminology and concepts used by Synopsys.

Starting from a brief description and purpose of some of the tools provided by Synopsys, the chapter covered the Synopsys environment that included examples of startup files needed, both for DC and PT, followed by the concepts of Objects, Variables and Attributes.

A brief introduction was also provided for the find command and its usefulness. Different Synopsys formats were discussed along with Design entry methods. The advantages and disadvantages of using read versus analyze/elaborate command were also covered.

Finally, the chapter concluded by describing some of the most useful directives used by DC for the purpose of hiding simulation only constructs.

Throughout the chapter, various examples were provided to facilitate the user in understanding these concepts.

4

SYNOPSYS TECHNOLOGY LIBRARY

Synopsys technology library format has almost become the de-facto library standard. Its compact yet informative format allows adequate representation of the deep sub-micron technologies. The popularity of the Synopsys library format is evident from the fact that most place and route tools provide a direct translation of the Synopsys libraries, with almost a one-to-one mapping between the timing models in Synopsys libraries, and the place and route timing models. A basic understanding of the library format and delay calculation methods is the key for successful synthesis.

Designers usually do not concern themselves with full details of the technology library as long as the library contains a variety of cells, each with different drive strengths. However, in order to optimize the design successfully, it is essential for designers to have a clear understanding of the delay calculation method used by DC along with the wire-load modeling and cell descriptions. It is therefore, the intent of this chapter to describe the Synopsys technology library from the designer's perspective, rather than discussing details about the structural and functional syntax of the library.

4.1 Library Basics

The Synopsys technology library is a text file (usually with extension ".lib"), which is compiled using the Library Compiler (LC) to generate a binary format with ".db" extension. The technology library contains the following information:

a) Library group
b) Library level attributes
c) Environment description
d) Cell description

4.1.1 Library Group

The library group statement specifies the name of the library, followed by an open brace. The closing brace is the last entry in the library file. Anything between the open and closing brace, is part of the entire library group description.

```
library (ex25) { /* start of library */
   ...
   < library description >
   ...
} /* end of library */
```

It is recommended that the file name and the technology library name be the same.

4.1.2 Library Level Attributes

The library level attributes are statements that apply to library as a whole. These generally contain library features such as technology type, date, revision, and default values that apply to the entire library.

SYNOPSYS TECHNOLOGY LIBRARY 65

```
library (ex25) {
    technology (cmos) ;
    delay_model              : table_lookup ;
    date                     : "Feb 29, 2000" ;
    revision                 : "1.0" ;
    current_unit             : "1A" ;
    time_unit                : "1ns" ;
    voltage_unit             : "1V" ;
    pulling_resistance_unit  : "1kohm" ;
    capacitive_load_unit (1.0, pf) ;
    default_inout_pin_cap    : 1.5 ;
    default_input_pin_cap    : 1.0 ;
    default_output_pin_cap   : 0.0 ;
    default_max_fanout       : 10.0 ;
    default_max_transition   : 3.0 ;
    default_operating_conditions : NOMINAL
    in_place_swap_mode       : match_footprint ;
    ……
    ……
}
```

4.1.3 Environment Description

Environment attributes are defined in the library to model the variations of temperature, voltage and manufacturing processes. These consist of scaling factors (derating), timing range models and operation conditions. In addition, the environment description also contains wire-load models that are used by DC to estimate interconnect wiring delays.

4.1.3.1 Scaling Factors

The scaling factors or K-factors are multipliers that provide means for derating the delay values based on the variations in process, voltage and temperature, or simply PVT. Only some of the K-factor statements are shown below as an example. Please refer to the library compiler reference manual for full details.

k_process_fall_transition	: 1.0 ;
k_process_rise_transition	: 1.2 ;
k_process_fall_propagation	: 0.4 ;
k_process_rise_propagation	: 0.4 ;
k_temp_fall_transition	: 0.03 ;
k_temp_rise_transition	: 0.04 ;
k_temp_fall_propagation	: 1.2 ;
k_temp_rise_propagation	: 1.24;
k_volt_fall_transition	: 0.02 ;
k_volt_rise_transition	: 0.5 ;
k_volt_fall_propagation	: 0.9 ;
k_volt_rise_propagation	: 0.85 ;

4.1.3.2 Operating Conditions

Sets of operating conditions defined in the library specify the process, temperature, voltage and the RC tree model. These are used during synthesis and timing analysis of the design. A library is characterized using one set of operating conditions. During synthesis or timing analysis, if another set of operating conditions is specified, then DC uses the K-factors to derate the delay values based upon the specified operating conditions. Library developers may define any number of operating conditions in the library. Typically the following operating conditions are defined in the technology library:

```
operating_conditions (WORST) {
        process     : 1.3 ;
        temperature : 100.0 ;
        voltage     : 2.75 ;
        tree_type   : worst_case_tree ;
}
operating_conditions (NOMINAL) {
        process     : 1.0 ;
        temperature : 25.0 ;
        voltage     : 3.00 ;
        tree_type   : balanced_tree ;
}
```

```
operating_conditions (BEST) {
        process       : 0.7 ;
        temperature   : 0.0 ;
        voltage       : 3.25 ;
        tree_type     : best_case_tree ;
}
```

The process, temperature and voltage attributes have already been explained previously. The **tree_type** attribute defines the environmental interconnect model to be used. DC uses the value of this attribute to select the appropriate formula while calculating interconnect delays. The **worst_case_tree** attribute models the extreme case when the load pin is at the most distant end of a net, from the driver. In this case the load pin incurs the full net capacitance and resistance. The **balanced_tree** model uses the case where all load pins are on separate and equal interconnect wires from the driver. The load pin in this case, incurs an equal portion of net capacitance and resistance. The **best_case_tree** models the case where the load pin is sitting right next to the driver. The load pin incurs only the net capacitance, without any net resistance.

4.1.3.3 Timing Range Models

The timing range models provide additional capability of computing arrival times of signals, based upon the specified operating conditions. This capability is provided by Synopsys to accommodate the fluctuations in operating conditions for which the design has been optimized. DC uses the timing ranges to evaluate the arrival times of the signals during timing analysis.

```
timing_range (BEST) {
        faster_factor  : 0.5 ;
        slower_factor  : 0.6 ;
}
timing_range (WORST) {
        faster_factor  : 1.2 ;
        slower_factor  : 1.3 ;
}
```

4.1.3.4 Wire-Load Models

The **wire_load** group contains information that DC utilizes to estimate interconnect wiring delays during the pre-layout phase of the design. Usually, several models appropriate to different sizes of the logic are included in the technology library. These models define the **capacitance**, **resistance** and **area** factors. In addition, the **wire_load** group also specifies **slope** and **fanout_length** for the logic under consideration.

The **capacitance**, **resistance** and **area** factors represent the wire resistance, capacitance and area respectively, per unit length of interconnect wire. The **fanout_length** attribute specifies values for the length of the wire associated with the number of fanouts. Along with fanout and length, this attribute may also contain values for other parameters, such as **average_capacitance**, **standard_deviation** and **number_of_nets**. These attributes and their values are written out automatically, when generating wire-load models through DC. For manual creation, only the values for fanout and length are needed, using the **fanout_length** attribute. For nets exceeding the longest length specified in the **fanout_length** attribute, the slope value is used to linearly interpolate the existing **fanout_length** value, in order to determine its value.

```
wire_load (SMALL) {
        resistance    : 0.2 ;
        capacitance   : 1.0 ;
        area          : 0 ;
        slope         : 0.5 ;
        fanout_length( 1, 0.020 ) ;
        fanout_length( 2, 0.042 ) ;
        fanout_length( 3, 0.064 ) ;
        fanout_length( 4, 0.087 ) ;
        . . . .
        fanout_length(1000, 20.0 ) ;
}
```

SYNOPSYS TECHNOLOGY LIBRARY

```
wire_load(MEDIUM) {
        resistance    : 0.2 ;
        capacitance   : 1.0 ;
        area          : 0 ;
        slope         : 1.0 ;
        fanout_length( 1, 0.022 ) ;
        fanout_length( 2, 0.046 ) ;
        fanout_length( 3, 0.070 ) ;
        fanout_length( 4, 0.095 ) ;
        . . . .
        fanout_length(1000, 30.0 ) ;
}
wire_load(LARGE) {
        resistance    : 0.2 ;
        capacitance   : 1.0 ;
        area          : 0 ;
        slope         : 1.5 ;
        fanout_length( 1, 0.025 ) ;
        fanout_length( 2, 0.053 ) ;
        fanout_length( 3, 0.080 ) ;
        fanout_length( 4, 0.110 ) ;
        . . . .
        fanout_length(1000, 40.0 ) ;
}
```

In addition to the wire_load groups, other attributes are defined in the library to automatically select the appropriate wire_load group, based on the total cell area of the logic under consideration.

```
wire_load_selection(AUTO_WL) {
    wire_load_from_area ( 0, 5000, "SMALL" ) ;
    wire_load_from_area ( 5000, 10000, "MEDIUM" ) ;
    wire_load_from_area (10000, 15000, "LARGE" ) ;
}
default_wire_load_selection  : AUTO_WL ;
default_wire_load_mode       : enclosed ;
```

It is recommended that the value of the default_wire_load_mode be set to "enclosed" or "segmented" instead of "top". The wire load modes and their application are described in detail in Chapter 6.

4.1.4 Cell Description

Each cell in the library contains a variety of attributes describing the function, timing and other information related to each cell. Rather than going into detail and describing all the attributes possible, only the relevant attributes and related information useful to designers are shown in the example below:

```
cell (BUFFD0) {
    area : 5.0 ;
    pin (Z) {
        max_capacitance : 2.2 ;
        max_fanout : 4.0 ;
        function : "I" ;
        direction : output ;
        timing () {
            . . . .
        }
        timing () {
            . . . .
        }
        related_pin : "I" ;
    }
    pin (I) {
        direction   : input ;
        capacitance : 0.04 ;
        fanout_load : 2.0 ;
        max_transition : 1.5 ;
    }
}
```

The **area** attribute defines the cell area as a floating-point number without any units followed by pin description and their related timing.

In addition, several design rule checking (DRC) attributes may be associated with each pin of the cell. These are:

- **fanout_load** attribute for input pins.
- **max_fanout** attribute for output pins.
- **max_transition** attribute for input or output pins.
- **max_capacitance** attribute for output or inout pins.

The DRC conditions are based on the vendor's process technology and should not be violated. The DRC attributes define the conditions in which the cells of the library operate safely. In other words, cells are characterized under certain conditions (output loading, input slope etc.). Designs violating these conditions may have a severe impact on the normal operation of the cells, thereby causing the fabricated chip to fail.

Even though, the previous example contains all four attributes, generally only two are used. In most cases, either the **fanout_load** along with **max_fanout**, or **max_transition** with **max_capacitance** are used.

The **fanout_load** and **max_fanout** DRC attributes are related to each other, in such that the **max_fanout** value at the output of the driver pin cannot exceed the sum of all **fanout_load** values at each input pin of the driven cells. Consider the cell (BUFFD0) shown in the previous example. This cell contains a **max_fanout** value of 4.0 associated to the output pin Z, while the **fanout_load** value at its input is 2.0. This cell therefore, cannot drive more than two of its own kind (BUFFD0) cells, since

max_fanout (4) = **fanout_load** (2) of 1st cell + **fanout_load** (2) of 2nd cell

If the DRC violations occur, then DC replaces the driving cell with another that has a higher **max_fanout** value.

The **max_transition** attribute is generally applied to the input pin, whereas the **max_capacitance** is applied to the output pin. Both attributes perform the similar function as the **max_fanout** and **fanout_load** attributes. The difference being, that the max_transition attribute defines that any net that has a transition time greater than the specified max_transition value of the load pin, cannot be connected to that pin. The max_capacitance at the output pin specifies that the output pin of the driver cell cannot connect to any net that has the total capacitance (interconnect and load pin capacitance) greater than, or equal to the maximum value defined at the output pin.

If DRC violations occur, then DC replaces the driving cell with another that has a higher max_capacitance value.

In addition, the output pin contains attributes defining the function of the pin, and the delay values related to the input pin. The input pin defines its' pin capacitance and the direction. The capacitance attribute should not be confused with the max_capacitance attribute. DC uses the capacitance attribute to perform delay calculations only, while the max_capacitance, as explained above, is used for design rule checking.

It is also worthwhile to mention here that for sequential cells, the clock input pin uses another attribute (clock : true) that specifies that the input pin is of type "clock". More details can be found in the Library Compiler Reference Manual.

The cell's DRC attributes are often the most criticized part of the cell library. Library developers often find it impossible to satisfy everyone and are often blamed for not implementing the "right" numbers for these attributes. The problem is caused because the library, to a certain extent is dependent upon the coding style and chosen methodology. What works perfectly for one design may produce inadequate results for another design. It is therefore the intent of this section to briefly explain the solutions that designers may use to tailor the library to suit their needs.

In order to accommodate the design requirements, it is possible to change the values of the above DRC attributes on a per cell basis. However, it must be noted that the DRC attributes set in the library can only be tightened, they cannot be loosened. This can only be done, if the attributes are pre-specified in the cell description. Users should realize that if these attributes are not already specified on the pin of the cell in the technology library, it is not be possible to add these attributes on the pin, from dc_shell.

For instance, to change the **max_fanout** value specified for pin Z of cell BUFFD0 (of library ex25 described previously), from 4.0 to 2.0; the following **dc_shell** command may be used:

```
dc_shell> set_attribute find(pin, ex25/BUFFD0/Z) max_fanout 2.0
```

One may also use wildcards in the above command to cover a variety of cells. This is useful for cases where a global change is required. For example, users may use the following command to change the **max_fanout** value on all cells with 0 drive strengths in the technology library:

```
dc_shell> set_attribute find(pin, ex25/*D0/Z) max_fanout 2.0
```

Similarly, the **set_attribute** command may be used to alter the value of other DRC attributes. The above command may be specified in the .synopsys_dc.setup file for global implementation.

4.2 Delay Calculation

Synopsys supports several delay models. These include the CMOS generic delay model, CMOS piecewise linear delay model and the CMOS non-linear table lookup model. Presently, the first two models are not in common use, due to their inefficiencies in representing the true delays caused by VDSM geometries. The non-linear delay model is the most prevalent delay model used in the ASIC world.

4.2.1 Delay Model

The non-linear delay model (NLDM) method uses a circuit simulator to characterize a cell's transistors with a variety of input slew rates, and output load capacitances. The results form a table, with input transition and output load capacitance as the deciding factor for calculating the resultant cell delay.

Figure 4-1, shown below depicts the resulting delays and slew rates, interpolated to produce a non-linear delay model. The model's accuracy depends on the precision and range, of the chosen input slew rates and load capacitances.

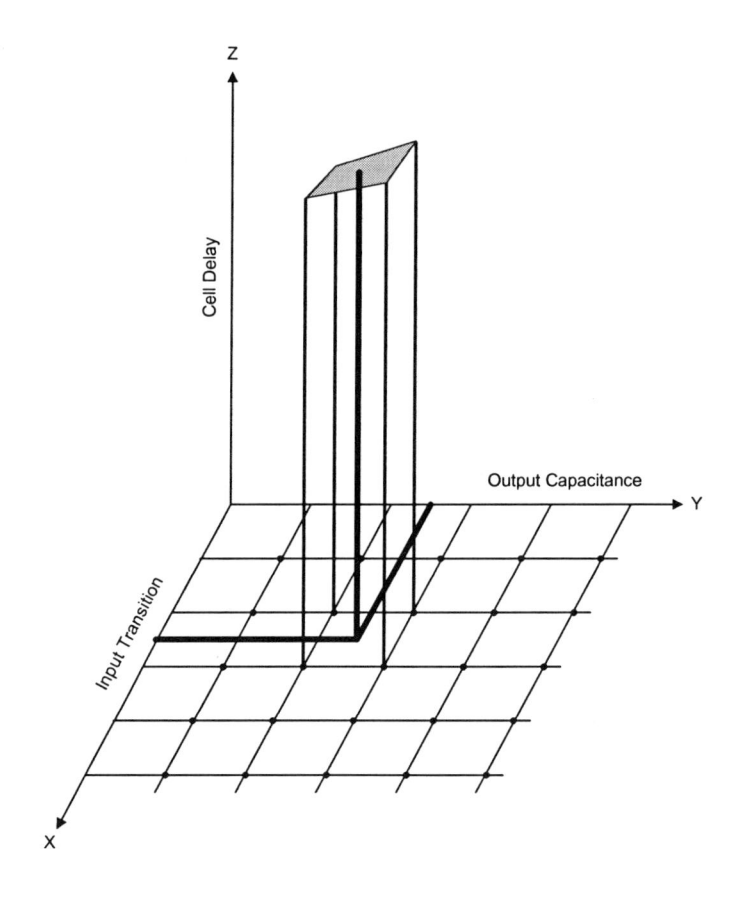

Figure 4-1. NLDM Table

If the delay number falls within the square (table in the library), then the delay is computed using interpolation techniques. The values of the surrounding four points are used to determine the delay value, using numerical methods. The problem arises, when any of the parameters fall outside the table. DC is best designed to extrapolate the resulting delay, but often ends up with an extremely high value. This may be a blessing in disguise, since a high value is easily noticeable during static timing analysis, providing designers an opportunity to correct the situation.

4.2.2 Delay Calculation Problems

The delay calculation of a cell is performed using the input transition time and the capacitive loading seen at the output. The input transition time of a cell is evaluated based upon the transition delay of the driving cell (previous cell). If the driving cell contains more than one timing arc, then the worst transition time is used, as input to the driven cell. This directly impacts the static timing analysis and the generated SDF file for a design.

Consider the logic shown in Figure 4-2. The signals, *reset* and *signal_a* are inputs to the instance U1. Let's presume that the *reset* signal is non critical as compared to *signal_a*. The *reset* signal is a slow signal, therefore, the transition time of this signal is high as compared to *signal_a*. This causes two transition delays to be computed for cell U1 (2 ns from A to Z, and 0.3 ns from B to Z). When generating SDF, the two values will be written out separately as part of the cell delay, for the cell U1. However, the question now arises, which of the two values does DC use to compute the input transition time for cell U2? DC uses the worst (maximum) transition value of the preceding gate (U1) as the input transition time for the driven gate (U2). Since the transition time of *reset* signal is more compared to *signal_a*, the 2ns value will be used as input transition time for U2. This causes a large delay value to be computed for cell U2 (shaded cell).

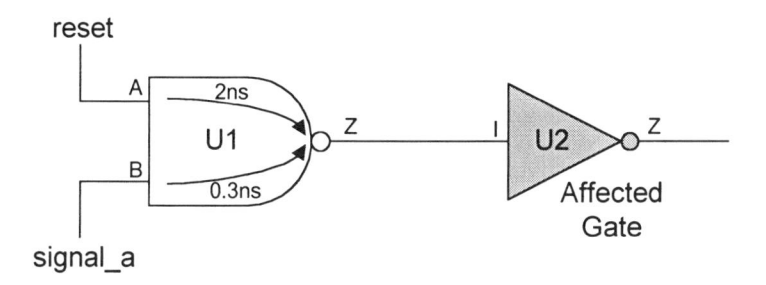

Figure 4-2. Delay Computation

To avoid this problem one needs to inform DC, not to perform the delay calculation for the timing arc – pin A to pin Z of cell U1. This step should be performed before writing out the SDF. The following dc_shell command may be used for this purpose:

```
dc_shell> set_disable_timing U1 -from A -to Z
```

Unfortunately, this problem also arises during static timing analysis. Failure to disable the timing computation of the false path leads to large delay values computed for the driven cell.

4.3 What is a Good Library?

Cell libraries determine the overall performance of the synthesized logic. A good cell library will result in fast design with smallest area, whereas a poor library will degrade the final result.

Historically, the cell libraries were schematic based. Designers would choose the appropriate cell and connect them manually to produce a netlist for the design. When the automatic synthesis engines became prevalent, the same schematic based libraries were converted and used for synthesis. However, since the synthesis engine relies on a number of factors for optimization, this approach almost always resulted in poor performance of the synthesized logic. It is therefore imperative that the cell library be designed catered solely towards the synthesis approach.

The following guidelines outline, the specific kind of cells in the technology library desired by the synthesis engine.

a) A variety of drive strengths for all cells.

b) Larger varieties of drive strengths for inverters and buffers.

c) Cells with balanced rise and fall delays (used for gated clocks).

d) Same logical function and its inversion as separate outputs, within the same physical cell (e.g., OR gate and NOR gate, as a single cell), again with a variety of drive strengths.

e) Same logical function and its inversion as separate cells (e.g., AND gate and NAND gate as two separate cells), with a variety of drive strengths.

f) Complex cells (e.g., AOI, OAI or NAND gate with one input inverted etc) with a variety of high drive strengths.

g) High fanin cells (e.g., AOI with 6 inputs and one output) with a range of different drive strengths.

h) Variety of flip-flops with different drive strengths, both positive and negative-edge triggered.

i) Single or Multiple outputs available for each flip-flop (e.g., Q only, or QN only, or both), each with a variety of drive strengths.

j) Flops to contain different inputs for Set and Reset (e.g., Set only, Reset only, no Set or Reset, both Set and Reset).

k) Variety of latches, both positive and negative-edge enabled each with different drive strengths.

l) Several delay cells. These are useful when fixing the hold-time violations.

Using the above guideline will result in a library optimized to handle the synthesis algorithm. This provides DC with the means to choose from a variety of cells to implement the best possible logic for the design.

It is worthwhile to note that the usage of high fanin cells, although useful in reducing the overall cell area, may cause routing congestion, which may inadvertently cause timing degradation, and/or increase in the area of the routed design. It is therefore recommended that these cells be used with caution.

Some designers prefer to exclude the low drive strengths for high fanin cells from the technology library. This is again is based on the algorithm used by the routing engine and the type of placement (timing driven etc.) used by designers. If the router is not constrained, then it uses a method by which it associates a weight to each net of the design while placing cells. Depending upon the weight of the net, the cells are pulled towards the source having the highest weight. High fanin cells have a larger weight associated to its inputs (because of the number of inputs) compared to the weight associated with their outputs (single output). Therefore, the router will place these cells near the gates that are driving it. This will result in the high fanin cell being pulled away from the cell it is supposed to be driving, causing a long net to be driven by the high fanin cell. If the high fanin cell is not strong enough to drive this long net (large capacitance) then the result will be the computation of large cell delay for the high fanin cell, as well as the driven gate (because of slow input transition time). By eliminating the low drive strengths of the high-fanin cells from the technology library, this problem can be prevented after layout.

4.4 Chapter Summary

To summarize, this chapter described the contents of the Synopsys technology library from the designer's perspective. The emphasis was placed upon the correct usage and understanding of the library, rather than focusing on details that are relevant only to library developers.

The chapter started with basics of the technology library, with separate groups within the library. The relevant portions of each group were explained in detail. This included explanation of all attributes that the library uses to perform its task.

Special emphasis was given to describing the delay calculation method, along with operating conditions, wire-load modeling and cell description. At each step, problems associated and workarounds were explained in detail.

Finally, suggestions were provided to the user as to what constitutes a good library optimized for synthesis engine. This includes helpful hints by taking into account the router behavior of the layout tool.

5

PARTITIONING AND CODING STYLES

Successful synthesis depends strongly on proper partitioning of the design, together with a good HDL coding style.

Logical partitioning is the key to successful synthesis (and place and route, if layout is hierarchical). Traditionally, designers partitioned the design in accordance with the functionality of each block, giving no thought to the synthesis process. As a result of incorrect partitioning, the inflexible boundaries degrade the synthesis results, which makes optimization difficult. Partitioning the design correctly can significantly enhance the synthesized result. In addition, reduced compile time and simplified script management is also achieved.

A good coding style is imperative, not only for the synthesis process, but also for easy readability of the HDL code. Today, many designers only stress verifying the functionality of the design. Driven by time restriction and/or lack of communication between the team members, designers do not have the luxury of carefully scrutinizing the HDL coding style. The fact remains, however, that a good coding style not only results in reduction of chip area and aids in top-level timing, but also produces faster logic.

5.1 Partitioning for Synthesis

Partitioning can be viewed as, utilizing the "Divide and Conquer" concept to reduce complex designs into simpler and manageable blocks. Promoting design reuse is one of the most significant advantages to partitioning the design.

Apart from the ease in meeting timing constraints for a properly partitioned design, it is also convenient to distribute and manage different blocks of the design between team members.

The following recommendations achieve best synthesis results and reduction in compile time.

a) Keep related combinational logic in the same module.

b) Partition for design reuse.

c) Separate modules according to their functionality.

d) Separate structural logic from random logic.

e) Limit a reasonable block size (perhaps, maximum of 10K gates per block)

f) Partition the top level (separate I/O Pads, Boundary Scan and core logic).

g) Do not add glue-logic at the top level.

h) Isolate state-machine from other logic.

i) Avoid multiple clocks within a block.

j) Isolate the block that is used for synchronizing multiple clocks.

k) WHILE PARTITIONING, THINK OF YOUR LAYOUT STYLE.

PARTITIONING AND CODING STYLES 81

The **group** and **ungroup** commands provide the designer with the capability of altering the partitions in DC, after the design hierarchy has already been defined by the previously written HDL code. Figure 5-1, illustrates such an action.

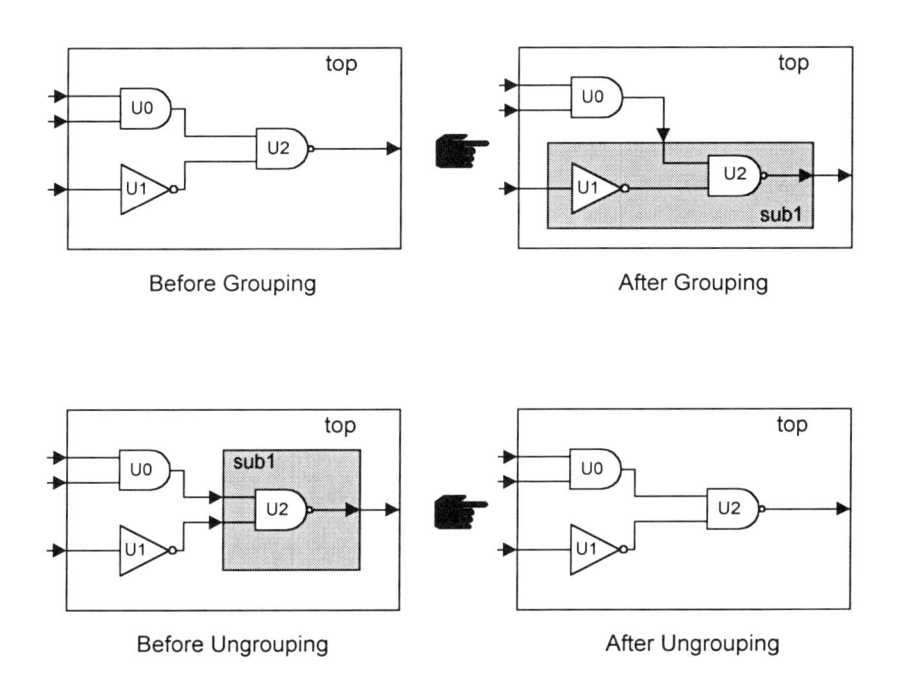

Figure 5-1. Changing Partitions

The **group** command combines the specified instances into a separate block. In Figure 5-1, instances U1 and U2 are grouped together to form a sub-block named sub1, using the following command.

```
dc_shell> current_design top

dc_shell> group {U1 U2} –design_name sub1
```

The ungroup command performs the opposite function. It is used to remove the hierarchy, as shown in Figure 5-1, by using the following command.

dc_shell> current_design top

dc_shell> ungroup –all

The designer can also use the ungroup command along with the –flatten and –all options to flatten the entire hierarchy. This is illustrated below:

dc_shell> ungroup –flatten –all

5.2   What is RTL?

Today, RTL or the Register Transfer Level is the most popular form of high-level design specification. An RTL description of a design describes the design in terms of transformation and transfer of logic from one register to another. Logic values are stored in registers where they are evaluated through some combinational logic, and then re-stored in the next register.

RTL functions like a bridge between the software and hardware. It is text with strong graphical connotations – text that implies graphics or structure. It can be described as technology independent, textual structural description, similar to a netlist.

5.2.1   Software versus Hardware

A frequent obstacle to writing HDL code is the software mind-set. HDLs have evolved from logic netlist representations. HDLs in their initial form (the Register Transfer Level) were a forum to represent logic in a format independent from any particular technology library. A higher level of HDL abstraction is the behavioral level that allows the design to be independent of timing and explicit sequencing. Lately, many system designers have adapted to HDLs to describe the full system.

Frequently, the expectation is that the synthesis tool will synthesize the HDL to the minimal area and maximum performance, regardless of how the HDL is written. The problem remains that at high level there are numerous ways of writing code to perform the same function. For example, a conditional expression could be written using *case* statements or *if* statements. Logically, these expressions are responsible for performing the same task, but when synthesized they can give drastically different results, as far as type of logic inferred, area, and timing are concerned. A reasonable caveat told to recent adopters of synthesis is – THINK HARDWARE!

5.3 General Guidelines

The following are general guidelines that every designer should be aware of. There is no fixed rule to adhere to these guidelines, however, following them vastly improves the performance of the synthesized logic, and produces a cleaner design that is well suited for automating the synthesis process.

5.3.1 Technology Independence

HDL should be written in a technology independent fashion. Hard-coded instances of library gates should be minimized. Preference should be given to inference rather than instantiation. The benefit being that the RTL code can be implemented with any ASIC library and new technology through re-synthesis. This is especially important for synthesizable IP cores that are commonly used by many designs.

In cases where placement of library gates is unavoidable, all the instantiated gates may be grouped together to form their own module. This helps in management of library specific aspects of a design.

5.3.2 Clock Logic

a) Clock logic including clock gating logic and reset generation should be kept in one block – to be synthesized once and not touched again. This helps in a clean specification of the clock constraints. Another advantage is that the modules that are being driven by the clock logic can be constrained using ideal clock specifications.

b) Avoid multiple clocks per block – try keeping one clock per block. Such restrictions later help avoid difficulties that may arise while constraining a block containing multiple clocks. It also helps in managing clock skew that may arise at the physical level. Sometimes this becomes unavoidable, for instance where synchronization logic is present to sync signals from one clock domain to the other. For such cases, it is recommended that designer isolate the sync logic, and synthesize it separately using special techniques. This includes setting a **dont_touch** attribute on the sync logic before instantiating it in the main block.

c) Clocks should be given meaningful names. A suggestion is to keep the name of the clock that reflects its functionality in addition to its frequency. Another good practice is to keep the same name for the clock, uniform throughout the hierarchy, i.e., the clock name should not change as it traverses through the hierarchy. This simplifies the script writing and helps in automating the synthesis process.

d) For DFT scan insertion, it is a requirement that the clocks be controlled from primary inputs. This may involve adding a mux at the clock source for controllability. Although not a hard and fast rule, it is recommended to hand-instantiate the clock-mux in the RTL (preferably at the top-level of the design). This allows the designer to know the instance name of the clock-mux at the stage of RTL elaboration, which in turn allows selective clock-mux timing path to be disabled, using **set_disable_timing**.

5.3.3 No Glue Logic at the Top

The top-level should only be used for connecting modules together. It should not contain any combinational glue logic. One of the benefits of this style is that it makes redundant the very time consuming top-level compile, which can now be simply stitched together without undergoing additional synthesis. Absence of glue logic at the top-level also facilitates layout, if performing hierarchical place and route.

5.3.4 Module Name Same as File Name

A good practice is to keep the module name (or entity name), same as the file name. Never describe more than one module or entity in a single file. A single file should only contain a single module/entity definition for synthesis. This has enormous benefits in defining a clean methodology using scripting languages like PERL, AWK etc.

5.3.5 Pads Separate from Core Logic

Divide the top-level into two separate blocks "pads" and "core". Pads are usually instantiated and not inferred, therefore it is preferred that they be kept separate from the core logic. This simplifies the setting of the dont_touch attribute on all the pads of the design, simultaneously. By keeping the pads in a separate block, we are isolating the library dependent part of RTL code.

5.3.6 Minimize Unnecessary Hierarchy

Do not create unnecessary hierarchy. Every hierarchy sets a boundary. Performance is degraded, if unnecessary hierarchies are created. This is because DC is unable to optimize efficiently across hierarchies. One may use the ungroup command to flatten the unwanted hierarchies, before compiling the design to achieve better results.

5.3.7 Register All Outputs

This is a well-known Synopsys recommendation. The outputs of a block should originate directly from registers. Although not always practical, this coding/design style simplifies constraint specification and also helps optimization. This style prevents combinational logic from spanning module boundaries. It also increases the effectiveness of the characterize-write-script synthesis methodology by preventing the pin-pong effect that is common to this type of compilation technique.

5.3.8 Guidelines for FSM Synthesis

The following guidelines are presented for writing finite state machines that may help in optimizing the logic:

a) State names should be described using "enumerated types" in VHDL, or "parameters" in Verilog.

b) Combinational logic for computing the next state should be in its own *process* or *always* block, separate from the state registers.

c) Implement the next-state combinational logic with a *case* statement.

5.4 Logic Inference

High-level Description Languages (HDLs) like VHDL and Verilog are front-ends to synthesis. HDLs allow a design to be represented in a technology independent fashion. However, synthesis imposes certain restrictions on the manner in which HDL description of a design is written. Not all HDL constructs can be synthesized. Not only that, synthesis expects HDLs to be coded in a specific way so as to get the desired results. We can say that synthesis is template driven – if the code is written using the templates that are understood and expected by the synthesis tool, then the results will be correct and predictable. The templates and other coding patterns for synthesis are called coding styles. For quality results it is imperative that designers possess a keen understanding of the coding styles, logic inferences, and the corresponding logic structures that DC generates.

5.4.1 Incomplete Sensitivity Lists

This is one of the most common mistakes made by designers. Incomplete sensitivity lists may cause simulation mismatches between the source RTL and the synthesized logic. DC issues a warning for signals that are present in the *process* or *always* block, but are absent from the sensitivity list. This is primarily a simulation problem since the process does not trigger when sensitized (because of the missing signal in the sensitivity list). The synthesized logic, however, is generally correct for blocks containing incomplete sensitivity lists.

Verilog Example

```
always @(weekend or go_to_beach or go_to_work)
begin
   if (weekend)
      action = go_to_beach
   else if (weekday)
      action = go_to_work;
```

VHDL Example

```
process (weekend, go_to_beach, go_to_work)
begin
   if (weekend) then
      action <= go_to_beach;
   elsif (weekday) then
      action <= go_to_work;
   end if;
end process;
```

The examples illustrated above do not contain the signal "weekday" in their sensitivity lists. The synthesized logic may still be accurate, however, during simulation the process will not trigger each time the signal "weekday" changes value. This may cause a mismatch between the simulation result of the source RTL and the synthesized logic.

5.4.2 Memory Element Inference

There are two types of memory elements – latches and flip-flops. Latches are level-sensitive memory elements, while flip-flops in general are edge-sensitive. Latches are transparent as long as the enable to the latch is active. At the time the latch is disabled, it holds the value present at the D input, at its Q output. Flip-flops on the other hand, respond to rising or falling edge of the clock.

Latches are simple devices, therefore they cover less area as compared to their counterparts, flip-flops. However, latches in general are more troublesome because their presence in a design makes DFT scan insertion difficult, although not impossible. It is also complicated to perform static timing analysis on designs containing latches, due to their ability of being transparent when enabled. For this reason, designers generally prefer flip-flops over latches.

The following sub-sections provide detailed information on how to avoid latches, as well as how to infer them, if desired.

5.4.2.1 Latch Inference

A latch is inferred when a conditional statement is incompletely specified. An *if* statement with a missing *else* part is an example of incompletely specified conditional. Here is an example, both in Verilog and VHDL:

Verilog Example

```
always @(weekend)
begin
    if (weekend)
        action <= go_to_beach;
end
```

VHDL Example

```
process (weekend)
begin
    if (weekend = '1') then
        action <= go_to_beach;
end process;
```

PARTITIONING AND CODING STYLES 89

The above statement will cause the DC to infer a latch enabled by a signal called "weekend". In the above example, "action" is not given any value when the signal "weekend" is 0. Always cover all the cases in order to avoid unintentional latch inference. This may be achieved by using an *else* statement, or using a *default* statement outside the *if* branch.

A latch may also get inferred from an incompletely specified *case* statement in Verilog.

```
`define sunny  2'b00
`define snowy  2'b01
`define windy  2'b10

wire [1:0] weather;

case (weather)
    sunny : action <= go_motorcycling;
    snowy : action <= go_skiing;
    windy : action <= go_paragliding;
endcase;
```

In the above case statement only 3 of the 4 possible values of "weather" are covered. This causes a latch to be inferred on the signal "action". Note, for the above example the Synopsys full_case directive may also be used to avoid the latch inference as explained in Chapter 3. The following example contains the *default* statement that provides the fourth condition, thereby preventing the latch inference.

```
case (weather)
    sunny : action <= go_motorcycling;
    snowy : action <= go_skiing;
    windy : action <= go_paragliding;
    default : action <= go_paragliding;
endcase;
```

VHDL does not allow incomplete case statements. This often means that the *others* clause must be used, consequently the above problem does not occur in VHDL. However, latches may still be inferred by VHDL, if a particular output signal is not assigned a value in each branch of the *case* statement. The inference being that outputs must be assigned a value in all branches to prevent latch inference in VHDL.

```
case (weather) is
    when sunny   => action <= go_motorcycling;
    when snowy   => action <= go_skiing;
    when windy   => action <= go_paragliding;
    when others  => null;
end case;
```

The above example, although containing the *others* clause will infer latches because the output signal "action" is not assigned a particular value in the *others* clause. To prevent this, all branches should be completely specified, as follows:

```
case (weather) is
    when sunny   => action <= go_motorcycling;
    when snowy   => action <= go_skiing;
    when windy   => action <= go_paragliding;
    when others  => action <= go_paragliding;
end case;
```

5.4.2.2 Register Inference

DC provides a wide variety of templates for register inference. This is to support different edge-types of the clock and reset mechanisms. A register is inferred, when there is an edge specified in the sensitivity list. The edge could be a positive edge or a negative edge.

5.4.2.2.1 Register Inference in Verilog
In Verilog, a register is inferred when an edge is specified in the sensitivity list of an *always* block. One register is inferred for each of the variables assigned in the *always* block. All variable assignments, not directly dependent on the clock-edge should be made in a separate *always* block, which does not have an edge specification in its sensitivity list.

PARTITIONING AND CODING STYLES 91

A plain and simple positive edge-triggered D flip-flop is inferred using the following template:

```
always @(posedge clk)
   reg_out <= data;
```

In order to infer registers with resets, the reset signal is added to the sensitivity list, with reset logic coded within the *always* block. Following is an example of a D flip-flop with an asynchronous reset:

```
always @(posedge clk or reset)
   if ( reset )
      reg_out <= 1'b0;
   else
      reg_out <= data;
```

Having a synchronous reset is a simple matter of removing the "reset" signal from the sensitivity list. In this case, since the block responds only to the clock edge, the reset is also, only recognized at the clock edge.

```
always @(posedge clk)
   if ( reset )
      reg_out <= 1'b0;
   else
      reg_out <= data;
```

Negative edge-triggered flop may be inferred by using the following template:

```
always @(negedge clk)
   reg_out <= data;
```

Absence of negative edge-triggered flop in the technology library will result in DC inferring a positive edge-triggered flop with an additional inverter to invert the clock signal.

5.4.2.2.2 Register Inference in VHDL

In VHDL a register is inferred when an edge is specified in the *process* body. The following example illustrates the VHDL template to infer a D flip-flop:

```
reg1: process (clk )
begin
    if ( clk'event and clk = '1' ) then
        reg_out <= data;
    end if;
end process Reg1;
```

DC does not infer latches for variables declared inside functions, since variables declared inside functions are reassigned each time the function is called.

Coding style template for registers with asynchronous and synchronous resets are similar in nature to that of Verilog templates, shown in previous section.

Negative edge-triggered flop may be inferred by using the following template:

```
reg1: process (clk )
begin
    if ( clk'event and clk = '0' ) then
        reg_out <= data;
    end if;
end process Reg1;
```

Absence of negative edge-triggered flop in the technology library will result in DC inferring a positive edge-triggered flop with an additional inverter to invert the clock signal.

5.4.3 Multiplexer Inference

Depending upon the design requirements, the HDL may be coded in different ways to infer a variety of architectures using muxes. These may comprise of a single mux with all inputs having the same delay to reach the output, or a priority encoder that uses a cascaded structure of muxes to prioritize the input signals. A mixture of the above techniques is also commonly used to place the late arriving signal closer to the output.

The correct use of *if* and *case* statements is a complex topic that is outside the scope of this chapter. There are application notes (from Synopsys) and other published materials currently available that explain the proper usage of these statements. It is therefore the intent of this chapter to refer the users to outside sources for this information. Only brief discussion is provided in this section.

5.4.3.1 Use *case* Statements for Muxes

In general, *if* statements are used for latch inferences and priority encoders, while *case* statements are used for implementing muxes. It is recommended to infer muxes exclusively through *case* statements. The *if* statements may be used for latch inferencing and priority encoders. They may also be effectively used to prioritize late arriving signals. This kind of prioritizing may be implementation dependent. It also limits reusability.

To prevent latch inference in *case* statements the *default* part of the *case* statement should always be specified. For example, in case of a state machine, the default action could be that all states covered by the *default* clause cause a jump to the "start" state. Having a *default* clause in the *case* statement is the preferred way to write *case* statements, since it makes the HDL independent of the synthesis tool. Using directives like full_case etc makes the code dependent on the synthesis tool.

If the default action is to assign don't-cares, then a difference in behavior between RTL simulation and synthesized result may occur. This is because, DC may optimize the don't-cares randomly causing the resulting logic to differ.

5.4.3.2 *if* versus *case* Statements – A Case of Priorities

Multiple *if* statements with multiple branches result in the creation of priority encoder structure.

```
always @(weather or go_to_work or go_to_beach)
begin
    if (weather[0]) action = go_to_work;
    if (weather[1]) action = go_to_beach;
end
```

In the above example, the signal "weather" is a two-bit input signal and is used to select the two inputs, "go_to_work" and "go_to_beach", with "action" as the output. When synthesized, the cascaded mux structure of the priority encoder is produced as shown in Figure 5-2.

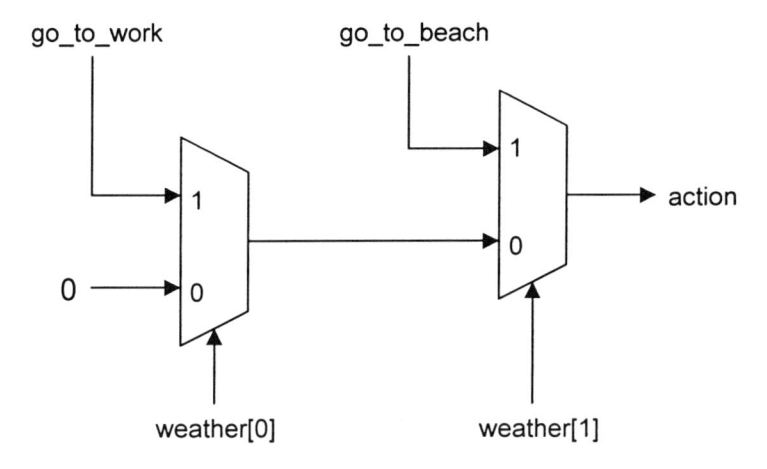

Figure 5-2. Result of using Multiple *if* Statements

If the above example used the *case* statement (instead of multiple *if* statements) in which all possible values of the selection index were covered and were exclusive, then it would have resulted in a single multiplexer as shown in Figure 5-3.

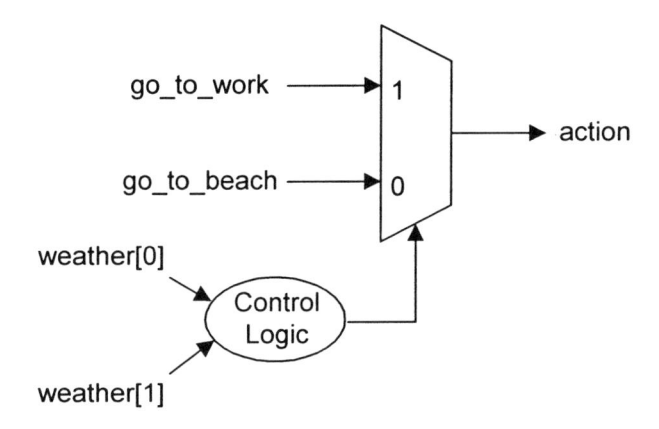

Figure 5-3. Result of using *case* Statement, or a Single *if* Statement

The same structure (Figure 5-3) is produced, if a single *if* statement is used, along with *elsif* statements to cover all possible branches.

5.4.4 Three-State Inference

Tri-state logic is inferred when high impedance (Z) is assigned to an output. Arbitrary use of tri-state logic is generally not recommended because of the following reasons:

a) Tri-state logic reduces testability.

b) Tri-state logic is difficult to optimize – since it cannot be buffered. This can lead to **max_fanout** violations and heavily loaded nets.

On the upside however, tri-state logic can provide significant savings in area.

Verilog

assign tri_out = enable ? tri_in : 1'bz;

VHDL

tri_out <= tri_in when (enable = '1') else 'Z';

5.5 Order Dependency

Both, Verilog and VHDL provide variable assignments that are order dependent/independent. Correct usage of these produces desired results, while incorrect usage may cause synthesized logic to behave differently than the source RTL.

5.5.1 Blocking versus Non-Blocking Assignments in Verilog

It is important to use non-blocking statements when doing sequential assignments like pipelining and modeling of several mutually exclusive data transfers. Use of blocking assignments within sequential processes may cause race conditions, because the final result depends on the order in which the assignments are evaluated. The non-blocking assignments are order independent; therefore they match closely to the behavior of the hardware.

Non-blocking assignment is done using the "<=" operator, while the "=" operator is used for blocking assignments.

```
always @(posedge clk)
begin
    firstReg    <= data;
    secondReg   <= firstReg;
    thirdReg    <= secondReg;
end
```

In hardware, the register updates will occur in the reverse order as shown above. The use of non-blocking assignments causes the assignments to occur in the same manner as hardware i.e., thirdReg will get updated with the old value of secondReg and the secondReg will get updated with the old value of firstReg. If blocking assignments were used in the above example, the signal "data" would have propagated all the way through to the thirdReg concurrently during simulation.

The blocking assignments should generally be used within the combinational *always* block.

5.5.2 Signals versus Variables in VHDL

Similar to Verilog, VHDL also provides order dependency through the use of signals and variables. The signal assignments may be equated to Verilog's non-blocking assignments, i.e., they are order independent. The variable assignments are order sensitive and correlate to Verilog's blocking assignments.

Variable assignments are done using the ":=" operator, whereas the "<=" operator is used for signal assignments.

The following example illustrates the usage of the signal assignments within the sequential *process* block. The resulting hardware contains three registers, with signal "data" propagating from firstReg to secondReg and then to the thirdReg. The RTL simulation will also show the same result.

```
process(clk)
begin
    if (clk'event and clk = '1') then
        firstReg    <= data;
        secondReg   <= firstReg;
        thirdReg    <= secondReg;
    end if;
end process;
```

A general recommendation is to only use signal assignments within sequential processes and variable assignments within the combinational processes.

5.6 Chapter Summary

This chapter highlighted the partitioning and coding styles suited for synthesis. Various guidelines and suggestions were provided to help the user code the RTL correctly with proper partitions, to make effective use of the synthesis engine.

The chapter began by suggestions on successful partitioning techniques and why they are necessary, followed by a short discussion on "what is RTL". Emphasis was given on "thinking hardware" while coding the design.

Next, general guidelines were covered that encompassed various suggestions and techniques, though not essential for synthesis, have significant impact on successful optimization. Adherence to these suggestions produce optimized designs that are well suited for automating the synthesis process.

An important section was devoted to the coding styles, and numerous examples were provided as templates to infer the correct logic. These included inference of latches, registers, multiplexers and three-state logic elements. At each step, advantages and disadvantages along with the correct usage was discussed.

The last section described the order dependency feature of both Verilog and VHDL languages. Also discussed were appropriate coding techniques to be used by utilizing the order dependency feature of both languages.

6

CONSTRAINING DESIGNS

This chapter discusses the process of specifying the design environment and its constraints. It describes various commonly used DC commands along with other helpful constraints that may be used to synthesize complex ASIC designs.

Please note that the commands described in this chapter only contain the most frequently used options. Designers are advised to consult the DC reference manual for the entire list of options available to a particular command.

This chapter contains information that is useful both for the novice and the advanced users of Synopsys tools. The chapter attempts to focus on "real world" applications, by taking into account deviations from the ideal situation. In other words, "Not all designs or designers, follow Synopsys recommendations". Incorporated within the chapter are numerous helpful ideas, marked as ☺ to guide the reader in real time application for selected commands.

6.1 Environment and Constraints

In order to obtain optimum results from DC, designers have to methodically constrain their designs by describing the design environment, target objectives and design rules. The constraints may contain timing and/or area information, usually derived from design specifications. DC uses these constraints to perform synthesis and tries to optimize the design with the aim of meeting target objectives.

6.1.1 Design Environment

Up till now, the assumption has been that the design has been partitioned, coded and simulated. The next step is to describe the design environment. This procedure entails defining for the design, the process parameters, I/O port attributes, and statistical wire-load models. Figure 6-1 illustrates the essential DC commands used to describe the design environment.

- **set_min_library** This is a new command, introduced in DC98 version. The command allows users to simultaneously specify the worst-case and the best-case libraries. This may be useful during initial compiles, preventing DC from violating the setup-time violations while fixing the hold-time violations.

    set_min_library <max library filename>
            –min_version <min library filename>

  dc_shell> set_min_library "ex25_worst.db"   \
            –min_version "ex25_best.db"

☺   The above command may be used for fixing hold-time violations during incremental compile or for in place optimization. In this case, the user should set both minimum and maximum values for the operating conditions.

CONSTRAINING DESIGNS

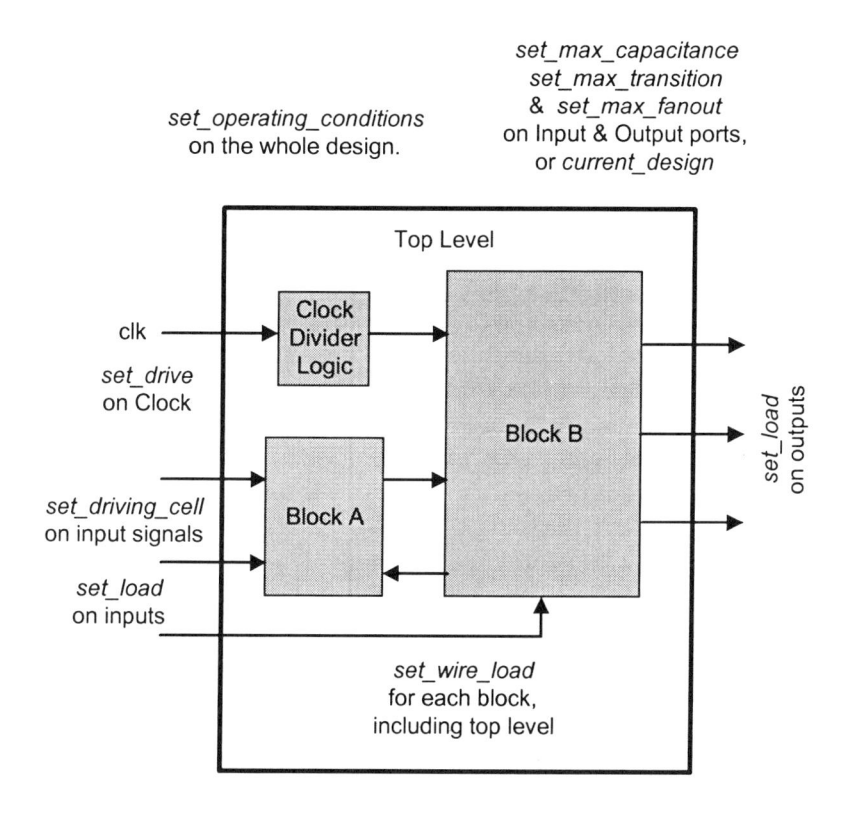

Figure 6-1. Basic Design Environment

- **set_operating_conditions** describes the process, voltage and temperature conditions of the design. The Synopsys library contains the description of these conditions, usually described as WORST, TYPICAL and BEST case. The names of operating conditions are library dependent. Users should check with their library vendor for correct setting. By changing the value of the operating condition command, full ranges of process variations are covered. The WORST case operating condition is generally used during pre-layout synthesis phase, thereby optimizing the design for maximum setup-time. The BEST case condition is commonly used to fix the hold-time violations. The TYPICAL case is mostly ignored, since analysis at WORST and BEST case also covers the TYPICAL case.

    set_operating_conditions <name of operating conditions>

dc_shell> set_operating_conditions WORST

☺ With the introduction of DC98, it is now possible to optimize the design both with the WORST and the BEST case, simultaneously. The optimization is achieved by using the −min and −max options in the above command, as illustrated below. This is very useful when optimizing the design to fix hold-time violations.

dc_shell> set_operating_conditions −min BEST −max WORST

− **set_wire_load** command is used to provide estimated statistical wire-load information to DC, which in turn, uses the wire-load information to model net delays as a function of loading. Generally, a number of wire-load models are present in the Synopsys technology library, each representing a particular size block. In addition, designers may also choose to create their own custom wire-load models to accurately model the net loading of their blocks.

    set_wire_load <wire-load model>
        −mode < top | enclosed | segmented >

dc_shell> set_wire_load MEDIUM −mode top

There are three modes associated for modeling wire-load models. These are **top**, **enclosed**, and **segmented**. Generally, only the first two modes are in common use. The **segmented** wire-load mode is not prevalent, since it relies on the wire-load models that are specific to the net segments.

The mode **top** defines that all nets in the hierarchy will inherit the same wire-load model as the top-level block. One may choose to use this wire-load model for sub-blocks, if planning to flatten them later for layout. This mode may also be chosen, if the user is synthesizing the design using the bottom-up compile method.

The second mode, **enclosed** specifies that all nets (of the sub-blocks) will inherit the wire-load model of the block that completely encloses the sub-blocks. For example, if the designer is synthesizing sub-blocks B and C that are completely enveloped by block A (which in turn is completely enclosed by the top-level), then sub-blocks B and C will inherit the wire-load models defined for block A.

The last mode, **segmented** is used for wires crossing hierarchical boundaries. In the above example, sub-blocks B and C will inherit the wire-load models specific to them, while the nets between sub-block B and C (but, within block A) will inherit the wire-load model specified for block A.

☺ It is extremely important that designers accurately model the wire loads of their design. Too optimistic or too pessimistic wire-load models result in increased synthesis iterations, in an effort to achieve timing convergence after post-layout. In general, during the pre-layout phase, slightly pessimistic wire-load models are used. This is done to provide extra timing margin that may get absorbed, by the routed design.

- **set_drive** and **set_driving_cell** are used at the input ports of the block. set_drive command is used to specify the drive strength at the input port. It is typically used to model the external drive resistance to the ports of the block or chip. The value of 0 signifies highest drive strength and is commonly utilized for clock ports. Conversely, set_driving_cell is used to model the drive resistance of the driving cell to the input ports. This command takes the name of the driving cell as its argument and applies all design rule constraints of the driving cell to the input ports of the block.

    set_drive <value> <object list>

    set_driving_cell –cell <cell name>
    　　　　　　　　–pin <pin name>
    　　　　　　　　<object list>

```
dc_shell> s t_drive 0 {CLK RST}

dc_shell> set_driving_cell –cell BUFF1 –pin Z all_inputs()
```

- **set_load** sets the capacitive load in the units defined in the technology library (usually pico farads, or pf), to the specified nets or ports of the design. It typically sets capacitive loading on output ports of the blocks during pre-layout synthesis, and on nets, for back-annotating the extracted post-layout capacitive information.

    set_load <value> <object list>

  ```
  dc_shell> set_load 1.5 all_outputs()

  dc_shell> set_load 0.3 find(net, "blockA/n1234")
  ```

- Design Rule Constraints or DRCs consist of **set_max_transition**, **set_max_fanout** and **set_max_capacitance** commands. These rules are generally set in the technology library and are determined by the process parameters. These rules should not be violated in order to achieve working silicon. Previous releases of DC (v97.08 and before) prioritized DRCs even at the expense of poor timing. However, the latest version DC98, prioritizes timing requirements over DRCs.

The DRC commands can be applied to input ports, output ports or on the current_design. Furthermore, if the value set in the technology library is not adequate or is too optimistic, then these commands may also be used at the command line, to control the buffering in the design.

    set_max_transition <value> <object list>

    set_max_capacitance <value> <object list>

    set_max_fanout <value> <object list>

CONSTRAINING DESIGNS 105

```
dc_shell> set_max_transition 0.3 current_design

dc_shell> set_max_capacitance 1.5 find(port, "out1")

dc_shell> set_max_fanout 3.0 all_outputs()
```

6.1.2 Design Constraints

Design constraints describe the goals for the design. They may consist of timing or area constraints. Depending on how the design is constrained, DC tries to meet the set objectives. It is imperative that designers specify realistic constraints, since unrealistic specification results in excess area, increased power and/or degradation in timing. The basic commands to constrain a design are shown in Figure 6-2.

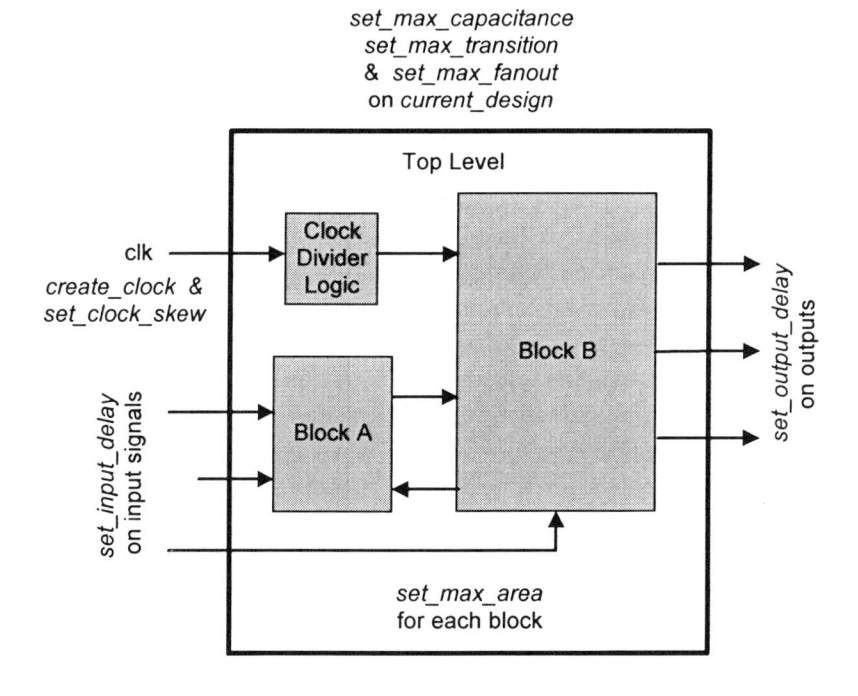

Figure 6-2. Basic Constraints for Synthesis

- **create_clock** command is used to define a clock object with a particular period and waveform. The –period option defines the clock period, while the –waveform option controls the duty cycle and the starting edge of the clock. This command is applied to a pin or port, object types.

Following example specifies that the port named CLK is of type "clock" that has a period of 40 ns, with 50% duty cycle. The positive edge of the clock starts at time, 0 ns, with the falling edge occurring at 20 ns. By changing the falling edge value, the duty cycle of the clock may be altered.

    dc_shell> create_clock –period 40 –waveform {0 20} CLK

☺ In some cases, a block may only contain combinational logic. To define delay constraints for this block, one can create a virtual clock and specify the input and output delays in relation to the virtual clock. To create a virtual clock, designers may replace the port name (CLK, in the above example) with the –name <virtual clock name>, in the above command. Alternatively, one can use the set_max_delay or set_min_delay commands to constrain such blocks. This is explained in detail in the next section.

- **set_dont_touch_network** is a very useful command, usually used for clock networks and resets. This command is used to set a dont_touch property on a port, or on the net. Note setting this property will also prevent DC from buffering the net, in order to meet DRCs. In addition, any gate coming in contact with the "dont_touched" net will also inherit the dont_touch attribute.

    dc_shell> set_dont_touch_network {CLK, RST}

☺ Suppose, you have a block that takes as input the primary clock, and generates secondary clocks e.g., clock divider logic. In this scenario, you should apply the set_dont_touch_network on the generated clock output port of the block. This will help prevent DC from buffering the clock network.

CONSTRAINING DESIGNS   107

☺ Caution should be exercised while using **set_dont_touch_network** command. For instance, if a design that contains gated clock circuitry and the set_dont_touch_network attribute has been applied to the clock input. This will prevent DC to appropriately buffer the gated logic, resulting in the DRC violation for the clock signal. The same will hold true for gated resets.

- **set_dont_touch** is used to set a dont_touch property on the current_design, cells, references or nets. This command is frequently used during hierarchical compilation of the blocks. Also, it can be used for, preventing DC from inferring certain types of cells present in the technology library.

   `dc_shell> set_dont_touch current_design`

   `dc_shell> set_dont_touch find(cell, "sub1")`

   `dc_shell> set_dont_touch find(net, "gated_rst")`

☺ For example, this command may be used on the block containing spare gates. The command will then instruct DC not to disturb (or optimize) the instantiation of the spare gates block.

- **set_dont_use** command is generally set in the .synopsys_dc.setup environment file. The command is instrumental in eliminating certain types of cells from the technology library that the user would not want DC to infer. For instance, by using the above command, you can filter out the flip-flops in your technology library whose name start with "SD" as illustrated below.

   `dc_shell> set_dont_use { mylib/SD* }`

- **set_input_delay** specifies the input arrival time of a signal in relation to the clock. It is used at the input ports, to specify the time it takes for the data to be stable after the clock edge. The timing specification of the design usually contains this information, as the setup/hold time requirements for input signals. Given the top-level timing specification of the design, this information may also be extracted for the sub-blocks of the design, by utilizing the top-down characterize compile method, explained in Chapter 7.

    dc_shell> set_input_delay –max 23.0 –clock CLK {datain}

    dc_shell> set_input_delay –min 0.0 –clock CLK {datain}

In Figure 6-3, the maximum input delay constraint of 23ns and the minimum input delay constraint of 0ns is specified for the signal *datain* with respect to the clock signal *CLK*, with a 50% duty cycle and a period of 30ns. In other words the setup-time requirement for the input signal *datain* is 7ns, while the hold-time requirement is 0ns.

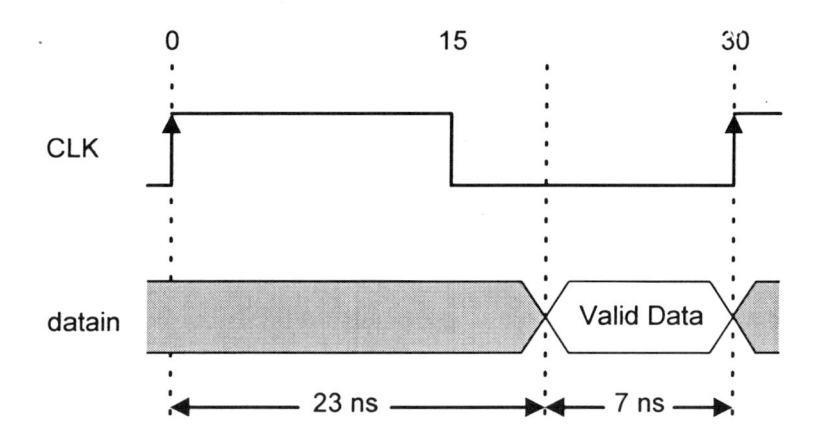

Figure 6-3. Specification of Input Delay

If both –min and –max options are omitted, the same value is used for both the maximum and minimum input delay specifications.

- **set_output_delay** command is used at the output port, to define the time it takes for the data to be available before the clock edge. The timing specification of the design usually contains this information. Given the top-level timing specification of the design, this information may also be extracted for the sub-blocks of the design, by utilizing the top-down characterize compile method, explained in Chapter 7.

    dc_shell> set_output_delay –max 19.0 –clock CLK {dataout}

In Figure 6-4, the output delay constraint of 19ns is specified for the signal *dataout* with respect to the clock signal *CLK*, with a 50% duty cycle and a period of 30ns. This means that the data is valid for 11ns after the clock edge.

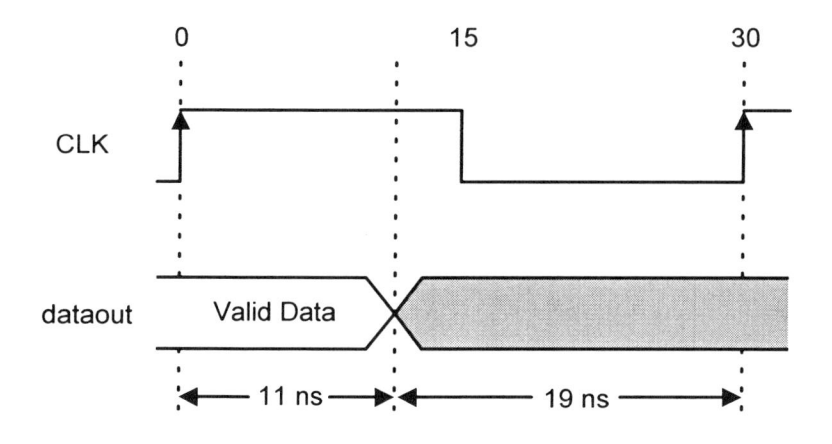

Figure 6-4. Specification of Output Delay

☺ During the pre-layout phase, it is sometimes necessary to over-constrain selective signals by a small amount to maximize the setup-time, thereby squeezing extra timing margin in order to reduce the synthesis-layout iterations. To achieve this, one may fool DC by specifying the over-constrained values to the above commands. Remember that over-constraining designs by a large amount will result in unnecessary increase in area and increased power consumption.

☺ A negative value (e.g., –0.5) may also be used to provide extra timing margin while fixing the hold-time violations after layout, by making use of the in-place optimization on the design, explained in Chapter 9.

- **set_clock_skew** command lets the user define the clock network delay and clock skew information. There are different options available for this command that facilitate the delay specification, both during the pre and the post-layout phases, as explained in the "Clocking Issues" section.

    `dc_shell> set_clock_skew 2.5 {CLK}`

- **set_clock_transition** for some reason does not get as much attention as it deserves. However, this is a very useful command, used during the pre-layout synthesis, and for timing analysis. Using this command forces DC to use the specified transition value (that is fixed) for the clock port or pin.

    `dc_shell> set_clock_transition 0.3 {CLK}`

☺ Setting a fixed value for transition time of the clock signal in pre-layout is essential because of a large fanout associated with the clock network. Using this command enables DC to calculate realistic delays for the logic being fed by the clock net based on the specified clock signal transition value. This is further explained in the "clocking issues" section later in the chapter.

6.2 Advanced Constraints

This section describes additional design constraints that go beyond the general constraints covered in the previous section. These constrains consist of specifying false paths, multicycle paths, max and min delays etc. In addition, this section also discusses the process of grouping timing critical paths for extra optimization.

It must be noted however, that the use of too many timing exceptions, such as false paths and multicycle paths causes significant impact on the run times.

- **set_false_path** is used to instruct DC to ignore a particular path for timing or optimization. Identification of false paths in a design is critical. Failure to do so, compels DC to optimize all paths in order to reduce total negative slack. Consequently, the critical timing paths may be adversely affected due to optimization of all the paths, which also includes the false paths.

    The valid startpoint and endpoint to be used for this command are the input ports or the clock pins of the sequential elements, and the output ports or the data pins of the sequential cells. In addition, one can further target a particular path using the -through switch.

    ```
    dc_shell> set_false_path –from in1 –through U1/Z –to out1
    ```

    ☺ Use this command when the timing critical logic is failing the static timing analysis because of the false paths.

- **set_multicycle_path** is used to inform DC regarding the number of clock cycles a particular path requires in order to reach its endpoint. DC automatically assumes that all paths are single cycle paths and will unnecessarily try to optimize the multicycle segment in order to achieve the timing. This may have a direct impact on adjacent paths as well as the area. Also, the command provides the –through option that facilitates isolating the multicycle segment in a design.

    ```
    dc_shell> set_multicycle_path 2  –from U1/Z     \
                                    –through U2/A  \
                                    –to out1
    ```

- **set_max_delay** defines the maximum delay required in terms of time units for a particular path. In general, it is used for the blocks that contain combinational logic only. However, it may also be used to constrain a block that is driven by multiple clocks, each with a different frequency. This command has precedence over DC derived timing requirements.

☺ For blocks, only containing combinational logic, one may either create a virtual clock and constrain the block accordingly, or use this command to constrain the total delay from all inputs to all outputs, as shown below:

   dc_shell> set_max_delay 5 –from all_inputs() –to all_outputs()

☺ Although, it is preferable to define only a single clock per block, there are situations where a block may contain multiple clocks, each with a different frequency. To constrain such a block, one may define all the clocks in the block using the normal **create_clock** and **set_dont_touch_network** commands. However, it becomes tedious to assign input delays of signals related to individual clocks. To avoid this situation, an alternative approach is to define the first clock (the most stringent one) using the normal approach, while constraining other clocks through the set_max_delay command, as shown below.

   dc_shell> set_max_delay 0 –from CK2   \
                              –to all_registers(clock_pin)

   The value of 0 signifies that a zero delay value is desired, between the input port CK2, and the input clock pins of all the flops within the block. In addition, one may also need to apply the set_dont_touch_network for other clocks. This method is suitable for designs containing gated clocks or resets.

- **set_min_delay** is the opposite of the set_max_delay command, and is used to define the minimum delay required in terms of time units for a particular path. Specifying this command in conjunction with the set_fix_hold command (described in Chapter 9) will instruct DC to add delays in the block to meet the minimum time unit specified. This command also has precedence over DC derived timing requirements.

```
dc_shell> set_min_delay 3 –from all_inputs() –to all_outputs()
```

- **group_path** command is used to bundle together timing critical paths in a design, for cost function calculations. Groups enable you to prioritize the grouped paths over others. Different options exist for this command, which include specification of critical range and weights.

  ```
  dc_shell> group_path –to {out1 out2} –name grp1
  ```

☺ Adding too many groups has significant impact on the compile time. Therefore, use it only as a last resort.

☺ Exercise caution while using this command. One may find that using this command increases the delay of the worst violating path, in the design. This is due to the fact, that DC prioritizes the grouped paths over other paths in the design. In order to improve the overall cost function, DC will try to optimize the grouped path over others and may degrade the timing of another group's worst violator.

6.3 Clocking Issues

In any design, the most critical part of synthesis is the clock description. There are always issues concerning the pre and post-layout definitions.

Traditionally in the past, big buffers were placed at the source of the clock to drive the full clock network. Thick clock spines were used in the layout for even distribution of clock network delays and to minimize clock skews. Although this method sufficed for sub-micron technologies, it is miserably failing in the VDSM realms. The interconnect RC's currently account for a major part of total delay. This is mainly due to the increase in resistance of the shrinking metal widths. It is difficult, if not impossible to model clocks using the traditional approach.

With the arrival of complex layout tools, capable of synthesizing clock trees, the traditional method has changed dramatically. Since, layout tools have the cell placement information, they are best equipped to synthesize the clock trees. It is therefore necessary to describe clocks in DC, such that it imitates the clock delays and skews of the final layout.

6.3.1 Pre-Layout

For reasons explained above, it is best to estimate the clock tree latency and skew during the pre-layout phase. To do this, use the following commands:

dc_shell> create_clock –period 40 –waveform {0 20} CLK

dc_shell> set_clock_skew –delay 2.5 –uncertainty 0.5 CLK

dc_shell> set_clock_transition 0.2 CLK

dc_shell> set_dont_touch_network CLK

dc_shell> set_drive 0 CLK

For the above example, a delay of 2.5 ns is specified as the overall latency for the clock signal CLK. In addition, the –uncertainty option approximates the clock skew. One can specify different numbers for the setup and hold uncertainties by using –minus_uncertainty for the setup-time, and –plus_uncertainty for the hold-time. For example:

dc_shell> set_clock_skew –delay 2.5   \
            –minus_uncertainty 2.0 –plus_uncertainty 0.2 CLK

Furthermore, specification of clock transition is essential. This restricts the max transition value of the clock signal. The delay through a cell is affected by the slope of the signal at its input pin and the capacitive loading present at the output pin. The clock network generally feeds large endpoints. This means, that although the clock latency value is fixed, the input transition time of the clock signal to the endpoint gates will still be slow. This results in DC calculating unrealistic delays (for the endpoint gates), even though in reality, the post-routed clock tree ensures fast transition times.

6.3.2 Post-Layout

Defining clocks after layout is relatively easy, since the user does not need to worry about the clock latency and skews. They are determined by the quality of the routed clock tree.

Some layout tools provide direct interface to DC. This provides a smooth mechanism for taking the routed netlist consisting of clock tree, back to DC. If this information is not present, then the user should extract the clock latency and the skew information from the layout tool. Using the pre-layout approach, this information can be used to define the clock latency and clock skew, as described before. If however, the netlist can be ported to DC, then the following commands may be used to define the clocks. For example:

```
dc_shell> create_clock –period 40 –waveform {0 20} CLK

dc_shell> set_clock_skew –propagated CLK            \
                        –minus_uncertainty 0.5      \
                        –plus_uncertainty 0.05  CLK

dc_shell> set_dont_touch_network CLK

dc_shell> set_drive 0 CLK
```

Notice the absence of the –delay option and inclusion of –propagated option. Since, we now have the clock tree inserted in the netlist, the user should propagate the clock instead of fixing it to a certain value. Similarly, the **set_clock_transition** command is no longer required, since DC will now calculate the input transition value of the clock network, based on the clock tree. In addition, a small clock uncertainty value may also be defined. This ensures a robust design that will function taking into account a wider process variance.

Some companies do not possess their own layout tools, but they rely on outside vendors to perform the layout. This situation of course varies from one company to the other. If the vendor provides the user, the post-routed netlist containing the clock tree, then the above method can be utilized. In some instances, instead of providing the post-routed netlist, the vendor only supplies the SDF file containing point-to-point timing for the entire clock network (and the design). In such a case, the user only needs to define the clock for the original netlist and back-annotate the SDF to the original netlist without propagating the clock. The clock skews and delays will be determined from the SDF file, when performing static timing analysis.

6.3.3 Generated Clocks

Many complex designs contain internally generated clocks. An example of this is the clock divider logic that may be used to generate secondary clock(s) of different frequency, derived from the primary clock source. If the primary clock has been designated as the clock source, then a limiting factor of DC is that does not automatically create a clock object for the generated clocks.

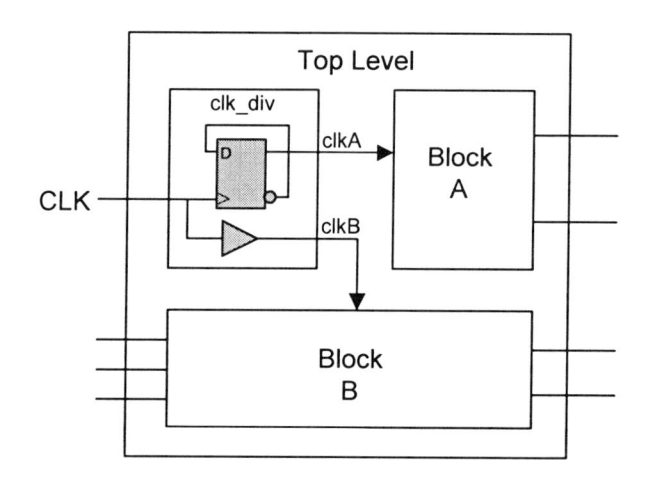

Figure 6-5. Specification of Generated Clock(s)

Consider the logic illustrated in Figure 6-5. A clock divider circuit in *clk_div* block, is used to divide the frequency of the primary clock *CLK* by half, and then generate the divided clock that drives *Block A*. The primary clock is also used to clock, *Block B* and is buffered internally (in the *clk_div* block), before feeding *Block B*.

CONSTRAINING DESIGNS

Assignment of clock object through create_clock command on *CLK* input to the top-level is sufficient for the clock feeding block B. This is because the *clkB* net inherits the clock object (through the buffer) specified at the primary source. However, *clkA* is not so fortunate. DC is unable to propagate the clock object throughout the entire net because the specification of clock object on primary source *CLK* stops at the register (shown as shaded flop). To avoid this situation, the clock object for *clkA* should be specified on the output port of the *clk_div* block. The following commands may be used to specify the clocks for the above example:

dc_shell> create_clock –period 40 –waveform {0 20} CLK

dc_shell> create_clock –period 80 –waveform {0 40}   \
                                        find(port, "clk_div/clkA")

6.4   Putting it Together

Example 6.1 provides a brief overview of some of the commands described in this chapter.

Example 6.1

```
/**********************************************/
/* Design entry */
    analyze –format verilog  sub1.v
    elaborate sub1

    analyze –format verilog  sub2.v
    elaborate sub2

    analyze –format verilog  top_block.v
    elaborate top_block

    current_design top_block
    uniquify
    check_design
```

```
/*********************************************/
/* Setup operating conditions, wire-load, clocks, resets */
    set_wire_load large_wl -mode enclosed
    set_operating_conditions WORST
    set_max_transition 1.0 top_block

    create_clock -period 40 -waveform {0 20} CLK
    set_clock_skew -delay 2.0  -minus_uncertainty 2.5  \
                                -plus_uncertainty 0.2  CLK
    set_dont_touch_network {CLK RESET}

/*********************************************/
/* Input drives */
    set_driving_cell -cell buff3 -pin Z all_inputs()
    set_drive 0 {CLK RST}

/*********************************************/
/* Output loads */
    set_load 0.5 all_outputs()

/*********************************************/
/* Set input & output delays */
    set_input_delay 10.0 -clock CLK all_inputs()
    set_input_delay -max  19.0 -clock CLK { IN1 IN2 }
    set_input_delay -min  -2.0 -clock CLK IN3

    set_output_delay 10.0 -clock CLK all_outputs()

/*********************************************/
/* Advanced constraints */
    group_path -from IN4 -to OUT2 -name grp1

    set_false_path -from IN5 -to sub1/dat_reg*

    set_multicycle_path 2 -from sub1/addr_reg/CP   \
                          -to sub2/mem_reg/D
```

```
/********************************************/
/* Compile and write the database */
   compile

   current_design  top_block

   write –hierarchy –output top_block.db
   write –format verilog –hierarchy –output top_block.sv

/********************************************/
/* Generate timing reports */
   report_timing –nworst 50
```

6.5   Chapter Summary

This chapter described all the basic and advanced commands used in DC, along with numerous tips to enhance the synthesis process. Focus was also given to the real time issues facing the designers as they descend deep into the VDSM technology.

A separate section was dedicated to issues related to clocks. This section described various techniques useful for specifying clocks, both for pre and post-layout. This section also included a topic on specification of generated clocks that are present in almost all designs. Finally, example DC scripts were included to guide the users to perform complex and successful synthesis.

7

OPTIMIZING DESIGNS

Ideally, a synthesized design that meets all timing requirements and occupies the smallest area is considered fully optimized. To achieve this goal, one must understand the behavior of synthesis process.

This chapter guides the reader to successfully optimize the design to obtain the best possible results.

7.1 Design Space Exploration

To achieve the smallest area while maximizing the speed of the design requires a fair amount of experimentation and iterative synthesis. The process of analyzing the design for speed and area to achieve the fastest logic with minimum area is termed – design space exploration.

Various factors influence the optimization process, primarily the coding style. While coding, designers generally focus on the functionality of the design and may not consider the synthesis guidelines, previously explained in Chapter 5 (This is a fact of life, we just have to live with it). At a later stage modifications to the HDL code are performed to facilitate the synthesis process. In reality, the HDL is generally fixed and only minor modifications are done, since major changes may impact other blocks or test benches. For this reason, changing the HDL code to help synthesis is less desirable.

For the sake of design space exploration, we can assume that the HDL code is frozen. It is now the designer's responsibility to minimize the area and meet the target timing requirements through synthesis and optimization.

Starting from version 98 of DC (or DC98) the previous compile flow changed. The timing is prioritized over area. This is shown in Figure 7-1. Another major difference between DC98 and previous versions is that, DC98 performs compilation to reduce "total negative slack" instead of "worst negative slack". This ability of DC98 produces better timing results but has some impact on the area. Also, in previous versions area minimization was handled automatically, however, DC98 requires designers to specify area constraints explicitly. Generally some area cleanup is performed by default even without specifying the area constraints but better results are obtained by including the constraints for area.

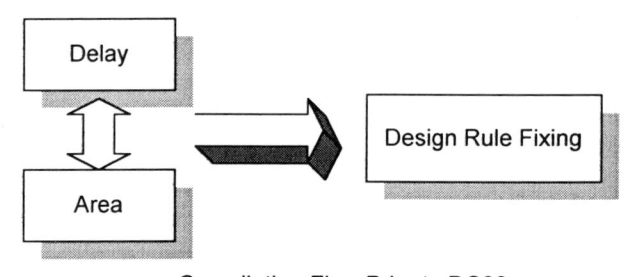

Compilation Flow Prior to DC98

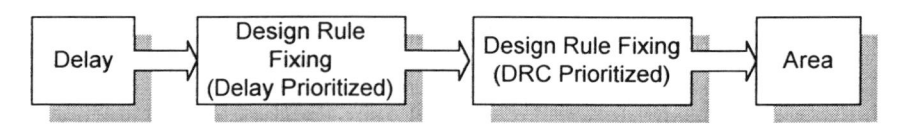

DC98 Compilation Flow

Figure 7-1. DC98 Changes

OPTIMIZING DESIGNS

Although, the delay is prioritized over area, it is extremely important to provide DC with realistic constraints. Some designers while performing bottom-up compile, fail to realize this point and over constrain the design. This causes DC to bloat the logic in order to meet the unrealistic timing goals. This is especially true for DC98 because it works on the reduction of total negative slack. This relationship between constraints and area is shown in Figure 7-2, which emphasizes that the area increases considerably with tightening constraints.

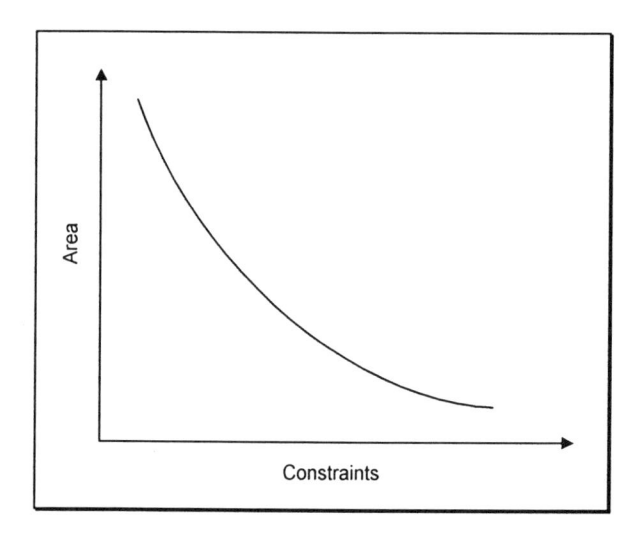

Figure 7-2. Constraints versus Area

Another representation of varying constraint is shown in Figure 7-3. This illustrates the relationship between constraints and delay across the design. It is shown that the actual delay of the logic decreases with tightening constraints, while relaxed constraints produce increased delay across the design. The horizontal part of the line on the left denotes that the constraints are so tight that further tightening of the constraints will not result in reduction of delay. Similarly the horizontal part of the line on the right signifies fully relaxed constraints, resulting in no further increase in delay.

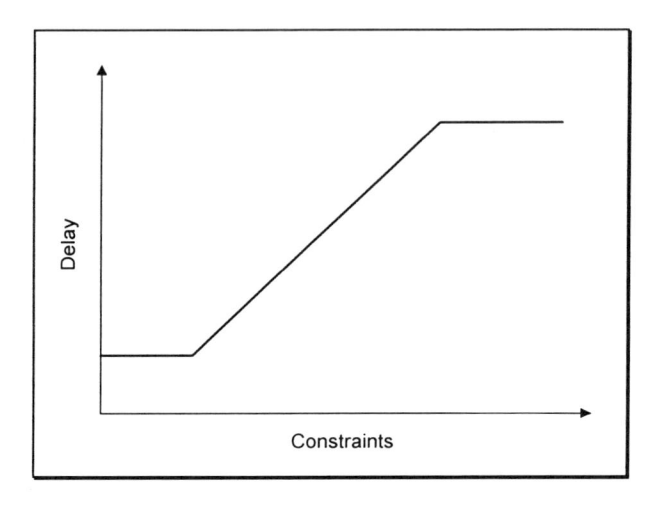

Figure 7-3. Constraints versus Delay

To further explain this concept, consider the diagram shown in Figure 7-4. For overly constrained design, DC tries to synthesize "vertical logic" to meet the tight timing constraints. However, if the timing constraints are non existent, the synthesized design would result in "horizontal logic", violating the actual timing specifications.

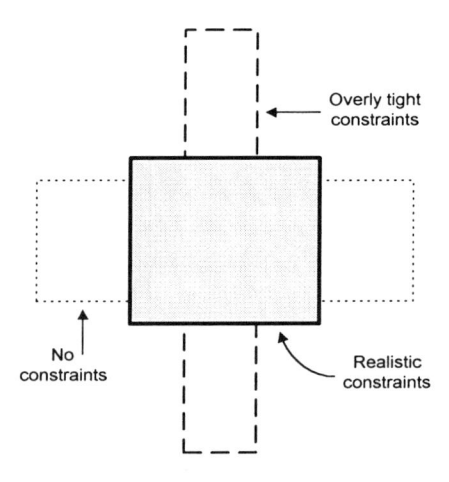

Figure 7-4. Horizontal versus Vertical Logic

The idea here is to find a common ground, by specifying realistic timing constraints. It is recommended to over constrain the design by a small amount (maybe 10 percent tighter than required) to avoid too many synthesis-layout iterations. This produces minimum area for the design, while still meeting the timing specifications. For this reason, choose the correct compile methodology, described in subsequent sections.

7.2 Total Negative Slack

The previous section briefly introduced the phrase "Total Negative Slack" or TNS for short. With the advent of DC98, a lot of importance has been given to this, and designers need to understand this concept to perform successful logic optimization.

Prior to DC98 version, DC would optimize the logic based on "Worst Negative Slack" or WNS. The WNS is defined as the timing violation (or negative slack) of a signal traversing from one startpoint to the endpoint for a particular path. During compile, DC would reduce the WNS one by one, in order to reduce total violations of the block. For this reason, grouping paths and specifying the critical range for timing-critical segments was considered essential.

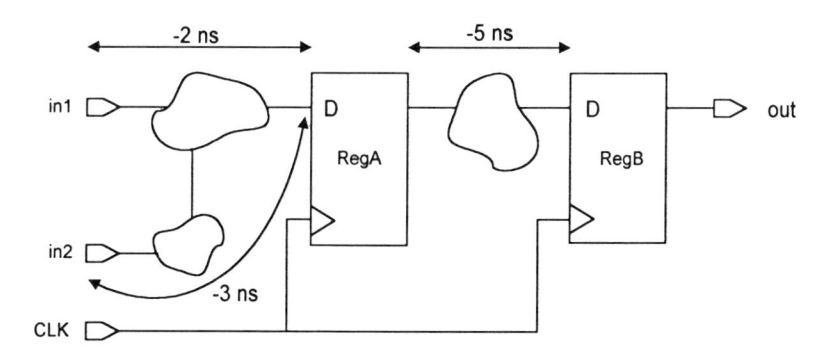

Figure 7-5. Total Negative Slack

DC98 not only, prioritizes delay over area but also targets TNS instead of WNS. To understand the concept of TNS consider the logic diagram shown in Figure 7-5. The WNS in this case is –5ns; from *RegA* to *RegB*. The TNS is the summation of all WNS per endpoint and in this case equals –8ns, i.e., WNS to *RegA* plus WNS to *RegB*.

There are several advantages to using this technique; primarily it produces lesser timing violations as compared to the previous method. Another benefit is, when using bottom-up compile methodology, the critical paths present at the sub-module may not be seen as critical from the top level. Reducing TNS of the overall design minimizes this effect. By providing smaller number of violating paths to the timing driven layout tool, less iterations between synthesis and layout can also be achieved.

Although, reduction of TNS over WNS produces less timing violations, it does have an impact on the overall area. It is recommended that you set area constraints, regardless of the kind of optimization performed. By default, DC98 prioritizes TNS over area. Area optimization occurs only for those paths with positive slack. In order to prioritize area over TNS, you may use the following command:

```
dc_shell> set_max_area 0 –ignore_tns
```

7.3 Compilation Strategies

Synopsys recommends the following compilation strategies that depend entirely on how your design is structured and defined. It is up to user discretion to choose the most suitable compilation strategy for a design.

a) Top-down hierarchical compile method.

b) Time-budget compile method.

c) Compile-characterize-write-script-recompile (CCWSR) method.

7.3.1 Top-Down Hierarchical Compile

Prior to the release of DC98, the top-down hierarchical compile method was generally used to synthesize very small designs (less than 10K gates). Using this method, the source is compiled by reading the entire design. Based on the design specifications, the constraints and attributes are applied, only at the top level. Although, this method provides an easy push-button approach to synthesis, it was extremely memory intensive and viable only for very small designs.

The release of DC98 provides Synopsys the capability to synthesize million gate designs by tackling much larger blocks (>100K) at a time. This indeed may be a feasible approach for some designs depending on the design style (single clock etc.) and other factors. You may use this technique to synthesize larger blocks at a time by grouping the sub-blocks together and flattening them to improve timing.

The advantages and disadvantages of this methodology are summarized below:

7.3.1.1 Advantages

a) Only top level constraints are needed.
b) Better results due to optimization across entire design.

7.3.1.2 Disadvantages

a) Long compile times (although, DC98 is much faster than previous releases).
b) Incremental changes to the sub-blocks require complete re-synthesis.
c) Does not perform well, if design contains multiple clocks or generated clocks.

128   *Chapter 7*

7.3.2   Time-Budgeting Compile

The second compilation approach to synthesis is termed as the time-budgeting strategy. This strategy is useful, if the design has been partitioned properly with timing specifications defined for each block of the design, i.e., designers have time budgeted the entire design, including the inter-block timing requirements.

The designer manually specifies the timing requirements for each block of the design, thereby producing multiple synthesis scripts for individual blocks. The synthesis is usually performed bottom-up i.e., starting at the lowest level and ascending to the topmost level of the design. This method targets medium to very large designs and does not require large amounts of memory.

Consider the following design, illustrated in Figure 7-6. The top level module incorporates blocks A and B. The specifications for both of these blocks are well defined and can be directly translated to timing constraints. For designs like these, the time-budgeting compilation strategy is ideal.

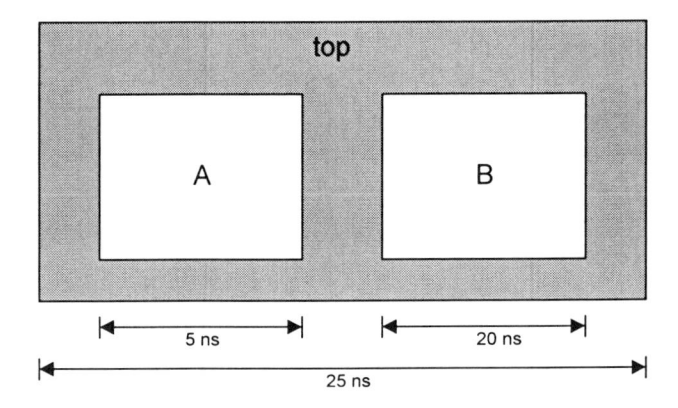

Figure 7-6. Design Suited for Time-Budgeting Compilation Strategy

OPTIMIZING DESIGNS 129

This advantages and disadvantages of this methodology are listed below:

7.3.2.1   Advantages

a) Easier to manage the design because of individual scripts.
b) Incremental changes to the sub-blocks do not require complete re-synthesis of the entire design.
c) Does not suffer from design style e.g., multiple and generated clocks are easily managed.
d) Good quality results in general because of flexibility in targeting and optimizing individual blocks.

7.3.2.2   Disadvantages

a) Tedious to update and maintain multiple scripts.
b) Critical paths seen at the top-level may not be critical at lower level.
c) The design may need to be incrementally compiled in order to fix the DRC's.

Figure 7-7 illustrates the directory structure and data organization, suited for this strategy. To automate the synthesis process, a makefile is used (refer to Appendix). The makefile specifies, the dependencies of each block and employs the user specified scripts (kept in the script directory) to compile the whole design, starting from the lowest level and ending at the top-most level. After the synthesis of each block, the results are automatically moved to their respective directories. The variables used in the makefile are defined in the users .cshrc file, for e.g., $SYNDB may be defined as: /home/project/design/syn/db

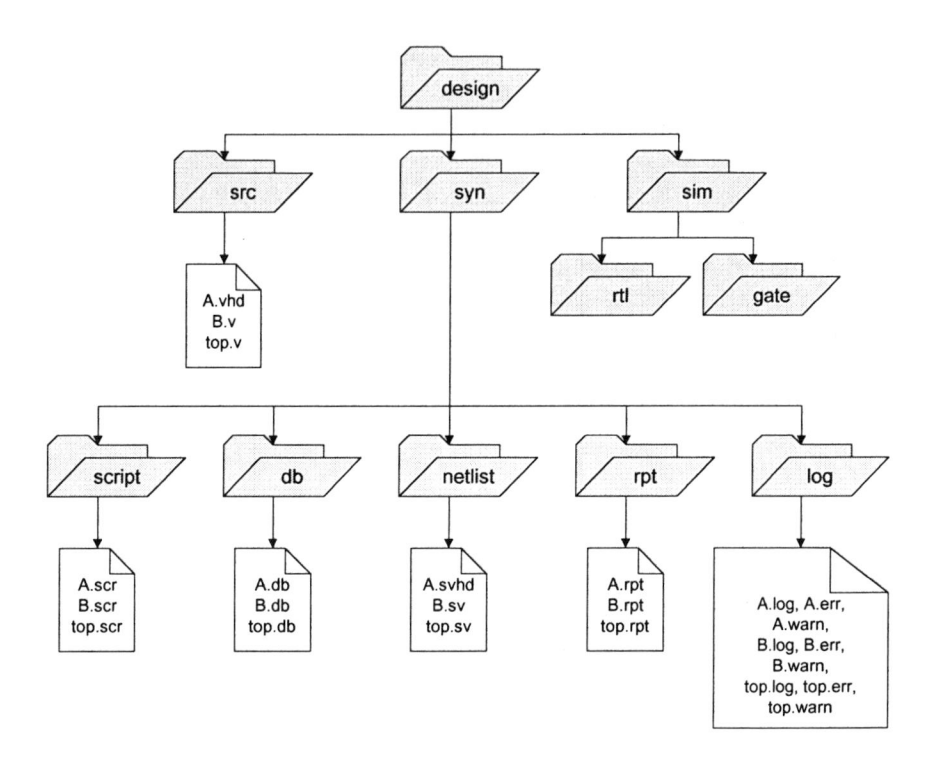

Figure 7-7. Directory Structure

7.3.3 Compile-Characterize-Write-Script-Recompile

The last compilation strategy covered here, is the Compile-characterize-write-script-recompile (CCWSR) method. This is an advanced synthesis approach, useful for medium to very large designs that do not have good inter-block specifications defined. This method is not limited by hardware memory and allows for time budgeting between the blocks.

This approach requires constraints to be applied at the top level of the design, with each sub-block compiled beforehand. The sub-blocks are then characterized using the top-level constraints. This in effect propagates the required timing information from the top-level to the sub-blocks. Performing a **write_script** on the characterized sub-blocks generates the constraint file for each sub-block. These constraint files are then used to re-compile each block of the design.

This approach normally produces the best results, since DC has been given the freedom of time budgeting between the sub-blocks of the design.

7.3.3.1 Advantages

a) Less memory intensive.
b) Good quality of results because of optimization between sub-blocks of the design.
c) Produces individual scripts, which may be modified by the user.

7.3.3.2 Disadvantages

a) The generated scripts are not easily readable.
b) Synthesis suffers from Ping-Pong effect. In other words it may be difficult to achieve convergence between blocks.
c) A change at lower level block generally requires complete re-synthesis of the entire design.

Example 7.1 illustrates the script that may be used to perform the CCWSR method.

Example 7.1   CCWSR synthesis script

```
/*********************************************************/
/* Setup environment */

SRC       = get_unix_variable("SRC")
all_modules = {A, B, top}
```

```
/**********************************************************/
/* Read the source and compile each module */

foreach (module, all_modules) {

        module_source = module + ".v"
        read –format verilog SRC + "/" + module_source
        current_design module
        link
        uniquify

    /* apply module attributes and constraints */
        module_script = module + ".scr"
        include  module_script

    /* compile each module */
        compile
        ungroup –flatten –all
        first_dbs = module + "_first.db"
        reset_design
        write –hierarchy –output  first_dbs
}

/**********************************************************/
/* Gather full design & specify top-level constraints */

foreach (module, all_modules) {
        first_dbs = module + "_first.db"
        read  first_dbs
}

read –format verilog SRC + "/" + top.v

current_design top
link
include top.scr    /** top-level constraints **/
write –hierarchy –output top_first.db
```

```
/******************************************************/
/* Characterize all instances in the design */

all_instances = {A, B, top}
characterize –constraint  all_instances

/******************************************************/
/* Generate scripts for each module */

foreach (module, all_modules) {
        current_design module
        char_module_script = module + ".wscr"
        write_script > char_module_script
}

/******************************************************/
/* Recompile each module */

foreach (module, all_modules) {
        remove_design –all

        module_source = module + ".v"
        read –format verilog SRC + "/" + module_source
        current_design module
        link
        uniquify

        char_module_script = module + ".wscr"
        include char_module_script

/******************************************************/
/* Recompile the module */

        compile
        module_db = module + ".db"
        write –hierarchy –output module_db
}
```

Usually, Some designers prefer the time-budgeting method, while others swear by the CCWSR method. However, for most designs the mix of these strategies provides the best results. Remember that it is possible to over constrain the design using the top-down compile or the time-budget method.

7.4 Resolving Multiple Instances

Before proceeding for optimization, one needs to resolve multiple instances of the sub-blocks of your design. This is a required step, since DC does not permit compilation until the multiple instances present in the design are resolved.

To better explain the concept of multiple instantiations of a module, consider the architecture of a design shown in Figure 7-8. Lets presume that you have chosen the time-budgeting compilation strategy and have synthesized moduleA separately. You are now compiling moduleB that instantiates moduleA twice as U1 and U2. The compilation will be aborted by DC with an error message stating that moduleA is instantiated 2 times in moduleB. There are two recommended methods of resolving this. You may either assign a dont_touch attribute to moduleA before reading moduleB, or uniquify moduleB. uniquify is a dc_shell command that in effect creates unique definitions of multiple instances. In this case, it will generate moduleA_U1 and moduleA_U2 (in Verilog), corresponding to instance U1 and U2 respectively as illustrated in Figure 7-9.

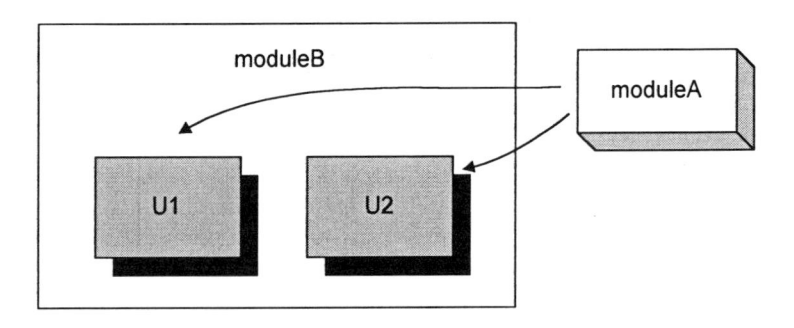

Figure 7-8. Non Uniquified Design

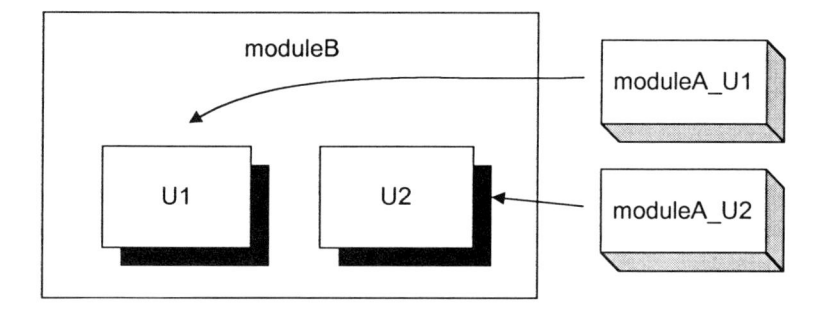

Figure 7-9. Uniquified Design

It is recommended to always uniquify the design, regardless of the compilation methodology chosen. The reason for this suggestion becomes evident while planning to perform clock tree synthesis during layout. This will be explained in detail in Chapter 9.

7.5 Optimization Techniques

This section describes various optimization techniques used to fine tune your design. Before we start on this subject, it is important to know that DC uses cost functions to optimize the design. This topic is thoroughly covered in the DC Reference manual, therefore will not be dealt with here. I would instead like to concentrate more on the practical optimization techniques instead of the mathematical algorithms used for calculating cost functions. Let's just say that DC calculates the cost functions based on the design constraints and DRCs to optimize the design.

7.5.1 Compiling the Design

Compilation of the design or modules is performed by the compile command. This command performs the actual mapping of the HDL code to gates from the specified target library. DC provides a range of options for this command, to fully control the mapping optimization of the design.

The command syntax along with most commonly used options is described below.

> compile –map_effort <low | medium | high>
> –incremental_mapping
> –in_place
> –no_design_rule | –only_design_rule
> –scan

By default, compile uses –map_effort medium, which usually produces ideal results for most of the designs. It also uses the default settings for structuring and flattening attributes, described in the next section. The "–map_effort high", should only be used, if the target objectives are not achieved through the default compile. This option enables DC to maximize its effort around the critical path by restructuring and re-mapping of logic, in order to meet the specified constraints. Beware, this usually produces long compile times.

The –incremental_mapping option is used, only after the initial compile (i.e., the design has been mapped to gates of the technology library), as it performs only at the gate level. This is a very useful and commonly used option. It is generally used to improve the timing of the logic and to fix DRCs. During incremental compile, DC performs various mapping optimizations in order to improve timing. Although Synopsys states that the resultant design will not worsen and may only improve in terms of design constraints; on rare occasions, using the above option may actually degrade the timing objectives. Users are therefore advised to experiment and use their own judgement. Nevertheless, the usefulness of this command is apparent while fixing DRCs at the top level of the design. To perform this, you may use –only_design_rule option while compiling incrementally. This prevents

OPTIMIZING DESIGNS 137

DC from performing mapping optimizations and concentrate only on fixing the DRCs.

The –no_design_rule option is not used frequently and as the name suggests, it instructs DC to refrain from fixing DRCs. You may use this option for initial passes of compile, when you don't want to waste time fixing DRC violations. At a later stage, generate the constraint report and then re-compile incrementally to fix DRCs. This is obviously a painful approach and users are advised to make their own judgement.

To achieve post layout timing convergence, it is sometimes necessary to resize the logic to fix timing violations. The –in_place option provides the capability of resizing the gates. Various switches available to designers to control the buffering of the logic govern this option. The usage of this option is described in detail in Chapter 9.

The –scan option uses the test ready compile feature of DC. This option instructs DC to map the design directly to the scan-flops – as opposed to synthesizing to normal flops before replacing them with their scan equivalents, in order to form the scan-chain. An advantage of using this feature is that, since the scan-flops normally have different timing associated with them as compared to their non scan equivalent flops (or normal flops), using this techniques makes DC take the scan-flop timing into account while synthesizing. This produces optimized scan inserted logic with correct timing.

7.5.2 Flattening and Structuring

Before we begin this discussion, it must be noted that the term "flattening" used here does not imply "removing the hierarchy". Flattening is a common academic term for reducing logic to a 2-level AND/OR representation. DC uses this approach to remove all intermediate variables and parenthesis (using boolean distributive laws) in order to optimize the design. This option is set to "false" by default.

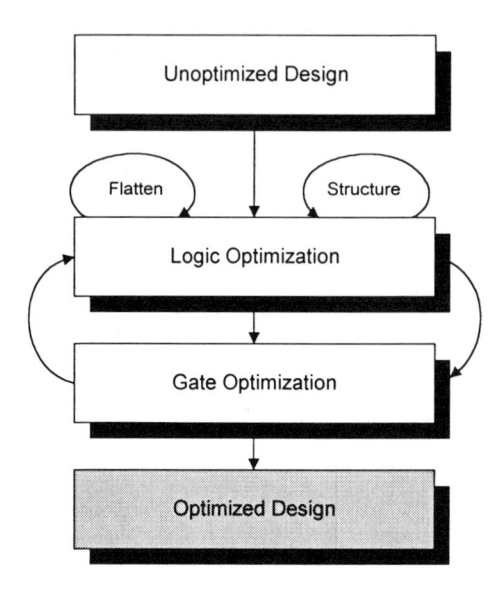

Figure 7-10. Optimization Steps

The optimization of the design is performed in two phases as shown in Figure 7-10. The logic optimization is performed initially by structuring and flattening the design. The resulting structure is then mapped to gates, using mapping optimization techniques. The default settings for *flatten* and *structure* attributes are:

Table 7-1. Default Settings for Flatten and Structure Attributes

Attribute	Value
flatten	false
structure	true
structure (timing)	true
structure (boolean)	false

As shown in the above table, flattening (set_flatten true) the design and Boolean optimization (set_structure –boolean true) is only performed when enabled.

7.5.2.1 Flattening

Flattening is useful for unstructured designs for e.g., random logic or control logic, since it removes intermediate variables and uses boolean distributive laws to remove all parenthesis. It is not suited for designs consisting of structured logic e.g., a carry-look-ahead adder or a multiplier.

Flattening results in a two-level, sum-of-products form, resulting in a vertical logic, i.e., few logic levels between the input and output. This generally results in achievement of faster logic, since the logic levels between the inputs and outputs are minimized. Depending upon the form of design flattened and the type of effort used, the flattened design can then be structured before the final technology mapping optimization. This is a recommended approach and should be performed to reduce the area, because flattening the design may cause a significant impact on the area of the design. A point to remember, if you flatten the design using "–map_effort high" option, DC may not be able to structure the design, therefore use this attribute judiciously.

In general, compile the design using default settings, since most of the time they perform adequately. Designs failing timing objectives may be flattened, with structuring performed as a second phase (on by default). If the design is still failing timing goals, then turn off structuring and flatten only. You may also experiment by inverting the phase assignment that sometimes produces remarkable results. This is done by setting the –phase option of the set_flatten command to "true". This enables DC to compare the logic produced by inverting the equation versus the non-inverted form of the equation.

For a hierarchical design, flatten attribute is set only on the current_design. All sub-blocks do not inherit this attribute. If you want to flatten the sub-blocks, then you have to explicitly specify using the –design option. The syntax for the flatten attribute along with most commonly used options is:

```
        set_flatten  <true | false>
                     –design <list of designs>
                     –effort <low | medium | high>
                     –phase <true | false>
```

7.5.2.2 Structuring

Structuring is used for designs containing regular structured logic, for e.g., a carry-look-ahead adder. It is enabled by default for timing only. When structuring, DC adds intermediate variables that can be factored out. This enables sharing of logic that in turn results in reduction of area. For example:

Before Structuring	After Structuring
P = a x + a y + c	P = a I + c
Q = x + y + z	Q = I + z
	I = x + y

It is important to note that, structuring produces shared logic that has an impact on the total delay of the logic. With the absence of specified timing constraints (or structuring is turned off with respect to timing), the logic produced will generally result in large delays across the block boundaries. Therefore, it is recommended that realistic constraints be specified, in addition to using the default settings.

Structuring comes in two flavors: timing (default) and boolean optimization. The latter is a useful method of reducing area, but has a greater impact on timing. Good candidates for boolean type of optimization are non critical timing circuitry e.g., random logic structures and finite state machines. As the name suggests, this algorithm uses boolean logic optimization to reduce area. Prior to version v1997.01, DC used a different algorithm to perform boolean optimization. Synopsys have since introduced another algorithm that is more efficient and requires less run time. This algorithm is based on automatic test pattern generation (ATPG) techniques to manipulate logic networks. To enable this algorithm, you have to set the following variable to "true":

    compile_new_boolean_structure = true

As with flattening, the set_structure command applies only to the current_design. The syntax of this command along with most commonly used options is:

```
set_structure  <true | false>
               –design <list of designs>
               –boolean <true | false>
               –timing <true | false>
```

In general, the design compiled with default settings produce satisfactory results. However, if your design is non-timing critical and you want to minimize for area only, then set the area constraints (set_max_area 0) and perform boolean optimization. For all other cases, structure with respect to timing only.

7.5.3 Removing Hierarchy

By default, DC maintains the original hierarchy of the design. The hierarchy is in-effect a logical boundary, which prevents DC from optimizing across this boundary. Many designers create unnecessary hierarchy for unknown reasons. This not only makes the synthesis process more cumbersome but also results in an increased number of synthesis scripts. As mentioned before, DC optimizes within logical boundaries. Having needless hierarchy in the design limits DC to optimize within that boundary without optimizing across the hierarchy.

Consider the logic shown in Figure 7-11(a). The top level (Block T) incorporates two blocks, A and B. The logic present at the output of block A and at the input of block B are separated by the block boundaries. Two separate optimizations of block A and B may not result in optimal solution. By combining block A and B (i.e., removing the boundaries) as shown in Figure 7-11(b), the two logic bubbles may be optimized as one, resulting in a more optimal solution. Designers (not Synopsys) refer to this process as "flattening" the design.

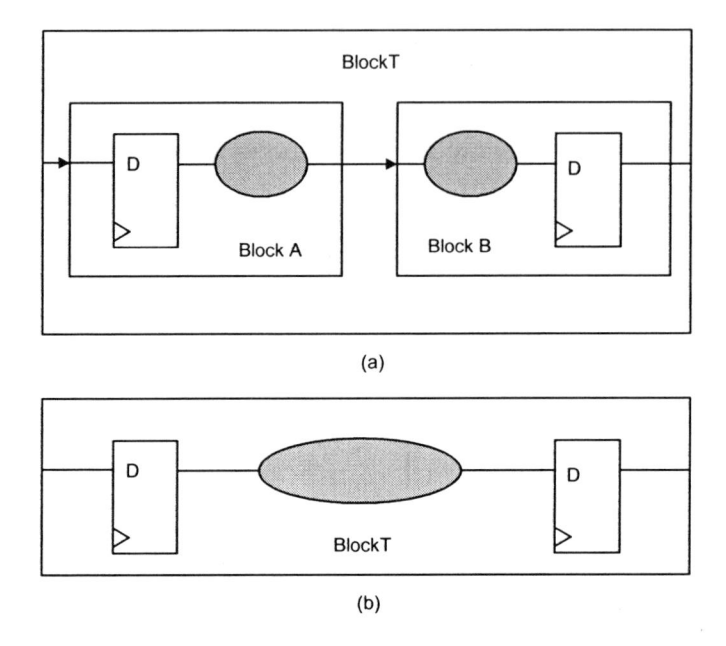

Figure 7-11. Ungrouping the Design to Improve Timing

To perform this, you may use the following command:

dc_shell> current_design BlockT

dc_shell> ungroup –flatten –all

7.5.4 Optimizing Clock Networks

Optimizing clock networks is one of the hardest operations to perform. This is due to the fact that as we descend towards VDSM technologies, the resistance of the metal increases dramatically causing enormous delays from the input of the clock pin to the registers. Also, low power design techniques require gating the clock to minimize the switching of the transistors when data is not needed to be clocked. This technique uses a gate (e.g., an AND gate), with inputs for clock and enable (used to enable or disable the clock source).

Previous methodologies included placement of a big buffer at the top level of the chip, near the clock source capable of driving all the registers in the design. Thick trunks and spines (e.g., fishbone structure) were used to fan the entire chip in order to reduce clock skew and minimize RC delays. Although this approach worked satisfactorily for technologies 0.5um and above, it is definitely not suited for VDSM technologies (0.35um and less). The above approach also meant increased number of synthesis-layout iterations.

With the advent of complex layout tools, it is now possible to synthesize the clock tree within the layout tool itself. The clock tree approach works best for VDSM technologies and although power consumption is a concern, the clock latency and skew are both minimal compared to the big buffer approach. The clock tree synthesis (CTS) is performed during layout after the placement of the cells and before routing. This enables the layout tool to know the exact placement location of the registers in the floorplan. It is then easy for the layout tool to place buffers optimally, so as to minimize clock skews. Since optimizing clocks are the major cause in increased synthesis-layout iterations, performing CTS during layout reduces this cycle.

We still have to optimize the clocks during synthesis before taking it to layout. We cannot assume that the layout tool will give us the magic clock tree that will solve all of our problems. Remember the more optimized your initial netlist, the better results you will get from the layout tool.

So how do we optimize clock networks during synthesis? By setting a set_dont_touch_network to the clock pin, you are assured that DC will not buffer up the network in order to fix DRCs. This approach works fine for most designs that do not contain clock-gating logic. But what if the clocks are gated? If you set the set_dont_touch_network on the clock that is gated then DC will not even size up the gate (let's assume a 2-input AND gate). This is because, the set_dont_touch_network propagates through all the combinational logic (AND gate, in this case), until it hits an endpoint (input clock pin of the register, in this case). This causes the combinational logic to inherit the dont_touch attribute also, which results in un-optimized gating logic that may violate DRCs, hence overall timing.

For instance, suppose the clock output from the AND gate is fanning out to a large number of registers and DC inferred a minimum drive strength for the AND gate. This will cause slow input transition times being fed to the registers resulting in horrendous delays for the clock net. To avoid this, you may remove the set_dont_touch_network attribute and perform incremental compilation. This will size up the AND gate and also insert additional buffers from the output of the AND gate, to the endpoints. Although, this approach seems ideal, it does suffer from some shortcomings. Firstly, it takes a long time for incremental compile to complete, and on rare occasions may produce sub-optimal results. Secondly, a lot of foresight is needed, for e.g., you need to apply set_dont_touch_network attribute on all other nets (resets and scan related signals that may not require clock tree).

A second approach is to find all high fanout nets in your design using the report_net command and buffer it from point to point using the balance_buffer command. (Refer to the DC reference manual for actual syntax for this command). Since, the balance_buffer command does not take clock skew into account, it should not be used as an alternative to clock tree synthesis.

Another technique is to perform in-place-optimization (IPO), using "compile –in_place", with compile_ok_to_buffer_during_inplace_opt switch, set to "false". This prevents DC from inserting additional buffers and will only size up the AND gate.

It must be noted that the above mentioned techniques are totally design dependant. Various methods have been provided that may be used for clock network optimization. Sometimes, you may find that you have to perform all the above methods to get optimal results and other times a single approach works perfectly.

Regardless of which method you use, you should also consider what you want to do during layout. For designs without gated clocks, it is preferable that CTS be performed at the layout level. For other design, with gated clocks, you have to analyze the clock in the design (pre and post-synthesis) carefully and take appropriate action. This may also include inserting the clock tree (during layout) after the AND gate for each bank of registers. Most layout tool vendors have realized this problem and offer various techniques to perform clock tree synthesis for gated clocks.

7.5.5 Optimizing for Area

By default, DC tries to optimize the design for timing. Designs that are non-timing critical but area intensive should be optimized for area. This can be done by initially compiling the design, with specification of area requirements, but no timing constraints. In other words, the design is synthesized with area requirements only. No timing constraints are used.

In addition, one may choose to eliminate the high-drive strength gates by assigning the dont_use attribute on them. The reason for eliminating high-drive strength gates is that they are normally used to speed up the logic in order to meet timing, however, they are larger in size. By eliminating their usage, considerable reduction in area may be achieved.

Once the design is mapped to gates, the timing and area constraints should again be specified (normal synthesis) and the design re-compiled incrementally. The incremental compile ensures that DC maintains the previous structure and does not bloat the logic unnecessarily.

7.6 Chapter Summary

Optimizing design, is the most time consuming and difficult task, since it depends enormously on various factors e.g., HDL coding styles, type of logic, constraints etc. This chapter described advanced optimization techniques and how they affect the synthesis process.

A detailed description of the impact on timing and area by varying design constraints is discussed. To reiterate, the best results are achieved by providing DC with realistic constraints. Over constraining a design results in large area and sub-optimal results.

With the introduction of DC98, the optimization flow has changed with more emphasis given on the timing, rather than area. Although by default area cleanup is always performed at the end of compilation, regardless, it is recommended that area constraints be specified. The timing optimization is performed by DC98 by minimizing the TNS. Prior to DC98, timing optimization was performed by reducing WNS per endpoint. The TNS optimization provides far superior results, although it does have an impact on the overall area of the design.

Various compile strategies are illustrated in this chapter, along with examples to automate this process. To successfully optimize a design, you may choose a single methodology or mix these strategies to get the desired result. All the strategies have their own advantages and disadvantages, which have also been illustrated. Choose the one, which best suits your design.

A separate section was devoted to uniquifying the design. Although, this step may not be needed, as argued by some designers, it is recommended to always uniquify the design because of reasons outlined in Chapter 9.

Finally, other optimization steps are discussed that emphasize on "how to produce optimal synthesized netlists". Various techniques, including clock network optimization and optimizing designs for area, were described along with recommended approaches.

8

DESIGN FOR TEST

The *Design-for-Test* or DFT techniques are increasingly gaining momentum among ASIC designers. These techniques provide measures to comprehensively test the manufactured device for quality and coverage.

Traditionally, testability was considered as an after thought, with implementation done only at the end of the design cycle. This approach usually provided minimal coverage and often led to unforeseen problems that resulted in increased cycle time. Merging testability features early in the design cycle was the final solution, creating the name *Design-for-Test*.

8.1 Types of DFT

Various vendors, including Synopsys provide solutions for incorporating testability in the design. Synopsys adds the DFT capabilities to DC through its Test Compiler (TC) that is incorporated within the DC suite of tools. The main DFT techniques that are currently in use today are:

a) Scan insertion
b) Memory BIST insertion
c) Boundary-Scan insertion

Of all three, scan insertion is the most complex and challenging technique, since it involves various design issues that need to be resolved, in order to get full coverage of the design.

8.1.1 Memory BIST

Unfortunately, Synopsys does not provide any solution for automatic memory BIST (*Built-In-Self-Test*) generation. Due to this reason memory BIST insertion is not covered in this section. However, there are vendors that do provide a complete solution, therefore a brief overview describing the main function of the memory BIST is included, providing designers an insight into this useful technique.

The memory BIST is comprised of controller logic that uses various algorithms to generate input patterns that are used to exercise the memory elements of a design (say a RAM). The BIST logic is automatically generated, based upon the size and configuration of the memory element. It is generally in the form of synthesizable Verilog or VHDL, which is inserted in the RTL source with hookups, leading to the memory elements. Upon triggering, the BIST logic generates input patterns that are based upon pre-defined algorithm, to fully examine the memory elements. The output result is fed back to the BIST logic, where a comparator is used to compare what went in, against what was read out. The output of the comparator generates a pass/fail signal that signifies the authenticity of the memory elements.

8.1.2 Boundary Scan DFT

JTAG or boundary scan is primarily used for testing the board connections, without unplugging the chip from the board. The JTAG controller and surrounding logic also may be generated directly by DC. Boundary-scan insertion is trivial, since the whole process is rather simple and mostly automatic. It is therefore the intent of this chapter to concentrate solely on the scan insertion techniques and issues. Readers are advised to refer to the Design Compiler Reference Manual for boundary scan insertion technique.

8.2 Scan Insertion

The scan insertion technique involves replacing all the flip-flops in the design, with special flops that contain built-in logic, solely for testability. The most prevalently used architecture is the *multiplexed* flip-flop. This type of architecture incorporates a 2-input mux at the input of the D-type flip-flop. The select line of the mux determines the mode of the device, i.e., it enables the mux to be either in the normal operational mode (functional mode with normal data going in) or in the test mode (with scanned data going in). These scan-flops are linked together (using the scan-data input of the mux) to form a scan-chain, that functions like a serial shift register. During scan mode, a combination of patterns are applied to the primary input, and shifted out through the scan-chain. If done correctly, this technique provides a very high coverage for all the combinational and sequential logic within a chip.

Other architectures available along with multiplexed type flip-flops are the *lssd* structure, *clocked scan* structure etc. As mentioned above, the most commonly used architecture is the *multiplexed flip-flop*. For this reason, throughout this section the focus remains on the *multiplexed flip-flop* type architecture, for DFT scan insertion.

8.2.1 Making Design Scannable

Synopsys provides designers the capability to perform scan insertion automatically, through its test-ready (or one-pass) compile feature. This technique allows designers to synthesize the design, and map the logic directly to the scan-flops, thus alleviating potential need for post-insertion adjustments.

Companies that do not use Synopsys tools for scan insertion, instead rely on other means to perform the same task. For such a case, replacing the normal flops in the synthesized netlist, with their scan equivalent flops, before linking the scan chains together, performs the scan insertion. It is strongly recommended that the static timing analysis should be performed again on the scan-inserted netlist, since some difference may exist between the characterized timing of the scan flops and their equivalent non-scan (normal) flops. This difference if not corrected may adversely affect the total slack of the design. To avoid this problem, library developers usually specify the scan-flops timing, to the normal-flops.

To enable the Synopsys test-ready compile feature, the scan style should be chosen prior to compilation. On a particular design, the set_scan_configuration command is used to instruct DC on how to implement scan. There are various options available for this command that may be used to control the scan implementation. Among others, these include options for clock mixing, number of scan chains and the scan style. Only the scan style option is illustrated below for the sake of clarity and explanation. Users are advised to refer to Design Compiler Reference Manual for syntax, and available range of options.

dc_shell> set_scan_configuration –style multiplexed_flip_flop

dc_shell> compile –scan

dc_shell> preview_scan

The compile –scan command compiles the design directly to scan-flops without linking them in a scan-chain, i.e., the scan insertion is not performed. The design is mapped to the scan-flops directly, instead of the normal flops. The design at this point is functionally correct, but un-scannable.

The preview_scan command is used to preview the scan architecture, specified through the set_scan_configuration command.

It is recommended that the check_test command be used after compilation, to check the design for testability related rule violations. DC flags any violations by issuing warnings/errors. Failure to fix these violations invariably results in reduced test coverage. The violations may occur due to various DFT related issues, encountered during scan insertion. Some of these issues and their solutions are discussed in the next section.

dc_shell> check_test

It is the designer's responsibility to correct the violations. This is mainly achieved by adding extra logic around the "problem" area to provide control

DESIGN FOR TEST

to the test logic. In order to fix these problems, modifying the source RTL instead of the netlist is the recommended approach. This approach allows the source RTL to remain as the "golden" database, that may be used for reference at some later stage. On the other hand, if the netlist is modified, then the changes may be forgotten, thus lost, after the design is taped-out.

Although, the scan has not yet been inserted, an additional step at this point is to get an estimate of the fault coverage of the design by generating the statistical ATPG test patterns. This step helps quantify the quality of the design, at an earlier stage. If the coverage number is low, then the only option is to identify and fix the areas that need further improvement. However, if the fault coverage number is high then this is an indication to proceed ahead.

It must be noted that the fault coverage numbers should be considered as best case only, due to the fact that the design may be part of a larger hierarchy, i.e., it may be a sub-block. At the sub-block level, the input port controllability and output port observability may be different when this sub-block is embedded in the full design (top-level). For a full design, this may cause a lower fault coverage number, than expected. The following command is used to generate the statistical test patterns:

dc_shell> create_test_patterns –sample < n >

Once the problem areas have been identified and fixed in the RTL, the design is ready for scan insertion. Using the following command performs the scan insertion:

dc_shell> insert_scan

The insert_scan command does more than just link the scan-flops together to form a scan-chain. It also disables the tri-states, builds and orders the scan-chains, and optimizes them to remove any DRCs. This command may insert additional test logic, in order to get better control over certain parts of the design.

After scan insertion the design should once again be checked for any rule violations through the check_test command. The report_test command may also be utilized to generate all test-related information about the design. Various options exist to control the output of the report. More details can be found in the Design Compiler Reference Manual.

8.2.2 Test Pattern Generation

Upon completion of scan insertion in the design, the test patterns may be generated for the entire design using the following command:

`dc_shell>` create_test_patterns

During dynamic simulation, the test patterns are used as input stimuli to the design to exercise all the scan paths. This step should be performed at the full chip level and preferably after layout.

8.3 DFT Guidelines

Obtaining high fault coverage for a design depends on the quality of the implemented DFT logic. Not all designs are ideal. Most "real-world" designs suffer from a variety of DFT related issues, and if left unsolved, result in reduced fault coverage. This section identifies some of these issues and provides solutions to overcome them.

8.3.1 Tri-State Bus Contention

This is one of the common problems faced by the DFT tool. During scan shifts, multiple drivers on a bus may drive the bus simultaneously, thus causing contention. Fixing this problem requires that only one driver be active at a given time. This can be achieved by adding the decoder logic in the design, which controls the *enable* input of each tri-state driver through a mux. The mux is used to select between the normal signal (in the functional mode) and the control line from the decoder. The decoder control is selected only during the scan-mode.

DESIGN FOR TEST

The decoder inputs are generally controlled directly from the primary inputs, thus providing means to selectively turn-on the tri-state drivers, thereby avoiding contention.

8.3.2 Latches

Avoid using latches as much as possible. Although, latches cover less area than flops, they are difficult to test. Testing maybe difficult but is not entirely impossible. Making them transparent during scan-mode can make them testable. This usually means, adding control logic (for the clock) to each latch. If an independent clock, clocks all the latches, then a single test-logic block may be used to control the clock to make the latches transparent during scan-mode.

8.3.3 Gated Reset or Preset

DFT requires that the reset/preset of a flop be controllable. If the reset/preset to a flop is functionally gated in the design, then the flop is un-scannable. To avoid this situation, the reset/preset signal should bypass the gating logic in scan-mode. A mux is generally used to remedy this problem, with the external scan-mode signal functioning as it's select line; and bypass reset/preset signal along with the original gated signal, as its input.

8.3.4 Gated or Generated Clocks

The gated clocks also suffer from the same issue that has been described above for gated resets. DFT requires that the clock input of the flop be controllable. The solution again is to bypass the gating logic through a mux, to make the flop controllable.

This problem is prevalent in those designs that contain logic to generate divided clocks. The flop(s) that is used to generate the divided clock should be bypassed during scan-mode. The dividing logic in this case may become un-scannable, but the divided clock can be controlled externally, thus providing coverage for the rest of the design. Small loss of coverage for the dividing logic is offset by the coverage gains achieved for the entire design.

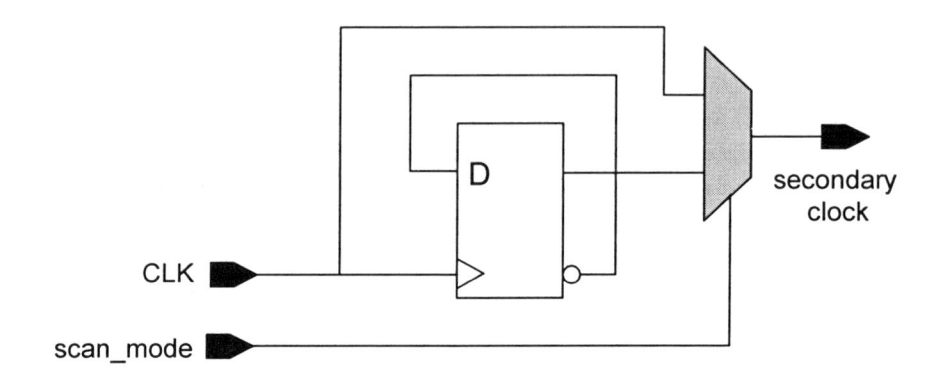

Figure 8-1. Bypassing Generated Clock

In Figure 8-1, the secondary clock is controlled externally, by using a mux that bypasses the CLK signal in the scan-mode. This provides controllability of the secondary clock, for the rest of the design. Depending upon the type of dividing logic being used, some parts of the logic may be un-scannable. The following command may be used to inform TC to exclude a list of sequential cells while inserting scan:

dc_shell> set_scan_element false <list of cells or designs>

8.3.5 Use Single Edge of the Clock

Most designs are coded using a single edge of the clock as reference. However, there may be situations where both the rising and falling edge of the clock is used. Although, this may have no impact on the timing of the functional data, it may create unnecessary timing problems for the scan data. The problem may be avoided by using the same clock edge for the entire design, when the design is in the scan-mode. This is illustrated in the following VHDL example:

```
process(clk, test_mode)
begin
    if (test_mode = '1') then
        muxed_clk_output <= clk;
    else
        muxed_clk_output <= not(clk);
    end if;
end process;
```

The above VHDL code infers a two-input mux. Positive edge of the clock is made use of during scan-mode, while the falling edge of the clock is used during normal mode.

8.3.6 Multiple Clock Domains

It is strongly recommended that designer assigns separate scan-chains for each clock domain. Intermixing of clock domains within a scan-chain typically leads to timing problems. This is attributed to the differences in clock skew between different clock domains. A disadvantage to using this technique is that it may lead to varying lengths of scan-chains.

An alternative solution is to group all flops belonging to a common clock domain, and connect them serially to form a single scan-chain. This requires the clock skew between the clock domain to be minimal. The clock sources should also be accessible from outside (primary inputs), so that the timing can be externally controlled when the device is tested at the tester.

There are other solutions available to this problem. One such solution is to use clock muxing at the clock source, so that only one clock is used during scan-mode.

8.3.7 Order Scan-Chains to Minimize Clock Skew

Presence of clock skew within a scan-chain usually causes hold-time violations. Some designers think that since testing is performed at slower speed as compared to the normal operational speed, the scan-chains cannot have any timing problems. This is a misconception. Only the setup-time is frequency dependent, while the hold-time is frequency independent. Therefore, it is extremely important to minimize clock skews to avoid any hold-time violations in the scan-chain.

The scan-chain may be re-ordered with flops having greater clock latency nearer to the source of the scan-chain, while the flops with less clock latency kept farther away. This helps in reducing the clock skew, thereby minimizing the possibility of any hold-time violations.

8.3.8 Logic Un-Scannable due to Memory Element

As explained earlier, the memory itself can be tested by the use of memory BIST circuitry. However, memory elements (e.g., RAMs) that do not have scan-chains (usually built-in) surrounding them, cause a loss of coverage for the combinational logic present at its inputs and outputs.

Figure 8-2, illustrates a case where a memory element is being fed by combinational logic. The logic present at its inputs is being shadowed by the RAM, thus is un-testable. If the inputs to the memory element are not coming directly from sequential elements, then any combinational logic present between the sequential logic and the memory element becomes un-testable. To avoid this situation, one may bypass the RAM in scan-mode. This is achieved by short-circuiting all the inputs feeding the RAM to the outputs of the RAM, through a mux. In scan-mode, the mux enables the short-circuited path and enables data to bypass the RAM.

Another problem that typically arises during scan-mode it that the outputs of the memory element are unknown. This typically results in the 'unknowns' being introduced to the surrounding scan-chain, causing it to fail. This situation can be avoided by using the bypass method, described above. The 'unknowns' generated by the RAM are blocked by the mux present at its outputs. This is because the mux is selected to bypass the RAM; it will therefore prevent the propagation of 'unknowns'.

DESIGN FOR TEST

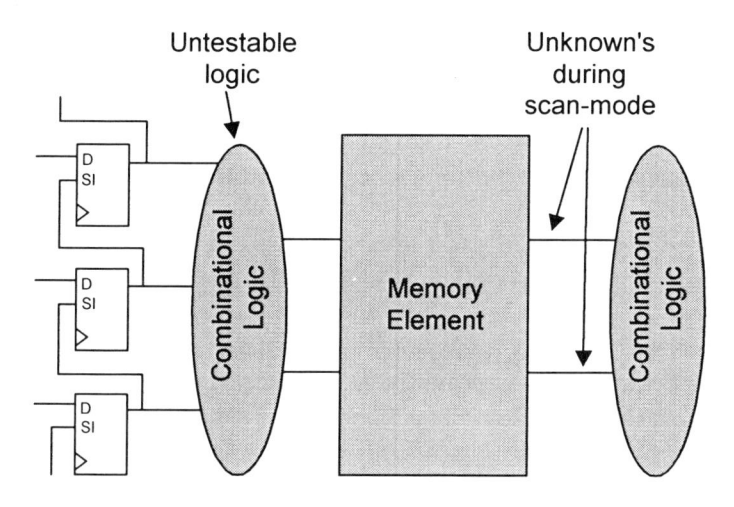

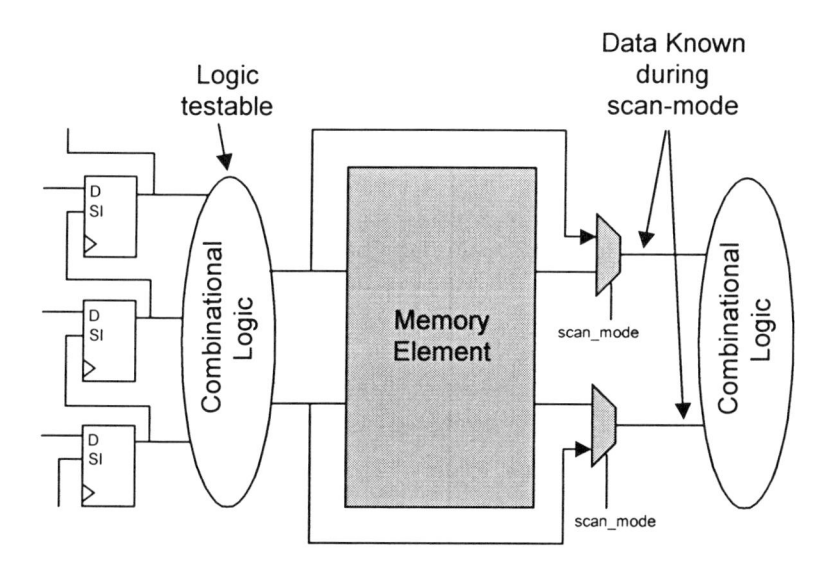

Figure 8-2. Memory Bypass

8.4 Chapter Summary

DFT techniques are essential to an efficient and successful testing of the manufactured device. By implementing DFT features early in the design cycle, full test coverage on the design may be achieved, thereby reducing the debugging time normally spent at the tester after the device is fabricated.

This chapter described the basic testability techniques that are currently in use, including a brief description of memory BIST that is not yet supported by Synopsys.

A detailed description was provided for the scan insertion DFT technique, using the Test Compiler. Various guidelines and solutions were also provided that may help the user to identify the issues and problems related to this technique.

9

LINKS TO LAYOUT AND POST LAYOUT OPTIMIZATION
Including Clock Tree Insertion

Until now, a virtual wall existed between the front-end and the back-end processes, with the signoff to the ASIC vendor for fabrication, occurring at the structural netlist level. The ASIC vendor was responsible for floorplanning and routing of the design, and provided the front-end designers the resulting delay data. However, this process was inefficient and often resulted in multiple exchanges of the netlist and the layout data between the designers and the ASIC vendor.

As we move deeper into the VDSM realm, the virtual wall between the front-end and the back-end is destined to collapse. This is because of the tremendous challenges and difficulties posed by the VDSM technologies. In order to overcome these difficulties, it is becoming evident that greater controllability and flexibility of the ASIC design flow is necessary. This requires total integration between the synthesis and layout processes. This means that designers are now compelled to perform their own layout. Instead of providing the ASIC vendors with the structural netlist, they are now given the physical database for final fabrication.

This shift in the signoff has resulted in a well-defined interface, between the synthesis tools and Place & Route tools (referred to as layout tools, from here onwards). Synopsys refers to this interface as links to layout or LTL.

This chapter describes the interface between DC and the layout tool. Almost all designs require the LTL interface to conduct the post-layout optimizations (PLO). Also, this chapter provides different strategies used for PLO. Furthermore, for successful layout, a section is devoted to clock tree synthesis, as performed by the layout tool.

Assuming that the user has synthesized and optimized a design. The design meets all timing and area requirements. Now the question arises, "How close are the estimated wire-load models used for pre-layout optimization, to the actual extracted data from the layout?" The only way to find this information is to floorplan and then route the design.

With shrinking geometries, the resistance of the wires is increasing as compared to its capacitance. This results in a large portion of the total delay (cell delay + interconnect delay) being dominated by the delays associated with the interconnect wires. In order to reduce this effect, designers are forced to spend an increased amount of time floorplanning the chip. Therefore, it is imperative for DC to make use of the physical information, in order to perform further optimizations.

Using LTL, one can exchange relevant data (e.g., timing constraints and/or placement information), to and from DC, to the layout tool. This helps DC perform improved post-layout optimizations. It also results in reduced iterations between synthesis and layout.

9.1 Generating Netlist for Layout

Most layout tools accept only the Verilog or EDIF netlist format as inputs. Many users, who code the design in VHDL, generate the netlist from DC in EDIF format for layout. Although this format is universal, it does have certain drawbacks. Primarily, the EDIF format is not easily readable; therefore modifying the netlist at a later stage to perform ECO is cumbersome. Secondly, the netlist in EDIF format is not simulatable.

So the question is, why should designers route a netlist that cannot be simulated? What happens if DC generates incorrect netlist (bad logic) due to some bug in its EDIF translator? With EDIF, the problem will only be identified at a much later stage, while performing LVS. Therefore, it is recommended that designers generate the netlist from DC in Verilog format, as input to the layout tool. Furthermore, the Verilog format is easy to understand, which considerably simplifies the task of modifying the netlist, in case an ECO needs to be performed on the design. In addition, even if the testbench for a design is in the VHDL format, one can still simulate the Verilog netlist by using simulators (currently available) that are capable of simulating a mixture of these languages.

Before sending the netlist (of the full design or individual block) to layout, it is recommended that the following procedure be performed on the netlist to facilitate smooth transfer of the design from DC to the layout tool.

a) Uniquify the netlist.

b) Simplify netlist by changing names of nets in the design.

c) Remove unconnected ports from the entire design.

d) Make sure that all pin names of leaf cells are visible.

e) Check for *assign* and *tran* statements.

f) Check for unintentional gating of clocks or resets.

g) Check for unresolved references.

9.1.1 Uniquify

As mentioned previously, the netlist must be uniquified in DC, in order to perform clock tree synthesis during layout. This operation generates a unique module/entity definition for a sub-block that is instantiated multiple times in the design. This may seem like an unnecessary operation that reduces the readability of the netlist, and results in increased size of the netlist. However, physically the design is considered flat by most layout tools. In other words, the blocks referenced multiple times, although ideal in all respects, exist physically at separate locations. Furthermore, flops present inside these blocks also need to be connected to the clock source. This makes it obvious that separate clock-net names are required for connecting the clock tree to these blocks.

Non-uniquified netlists pose a problem during clock tree transfer from the layout tool to DC. The problem only occurs, if the clock tree information alone is transferred to DC (through methods described later), that does not involve a complete netlist transfer from the layout tool to DC. In this case, only one module/entity definition for the multiple instanced blocks is present in the netlist, for a non-uniquified design. This causes a problem when the clock tree information is transferred back to DC, i.e., modifying the design database in DC, to include the buffers and additional ports in the sub-blocks. The problem is that two distinct net names (outputs of clock tree) cannot connect to the same port of a single module/entity. Uniquifying the design solves the above problem. However, it also causes the netlist to increase in size, since it creates separate module/entity definition for each instantiation of the block.

Some users prefer to uniquify the netlist as they traverse the hierarchy to reach the top-level, while others uniquify the whole chip at once, from the top-level. The recommended approach is to remove the dont_touch attribute from all sub-blocks of the design, before uniquifying the netlist.

The following command may be used to remove the dont_touch attribute from the entire design, before uniquifying the netlist from the top level:

dc_shell> remove_attribute find(−hierarchy design, "*") dont_touch

dc_shell> uniquify

9.1.2 Tailoring the Netlist for Layout

Some layout tools may have difficulty reading the Verilog netlist that contains unusual net names. For instance, DC sometimes produces signal names with "*cell*" or "*-return" appended, or in-between the names. Other times, users may find that some net names (or port names) have leading or trailing underscores. Also, most layout tools have restrictions on the maximum number of characters for a net, or a port name. Depending on the restrictions imposed by the specific layout tool, it is possible for the user to clean the netlist within DC, before writing it out. This ability of DC provides a smooth interface to the layout tool while meeting all tool requirements.

To prevent DC from generating the undesirable signal names, the user must first define rules, and then instruct DC to conform to these rules before writing out the netlist. For instance, one may define rules called "BORG" by including the following in the ".synopsys_dc.setup" file:

```
define_name_rules BORG  –allowed "A-Za-z0-9_"            \
                 –first_restricted "_" –last_restricted "_"   \
                 –max_length 30                                \
                 –map { {"\*cell\*", "mycell"}, {"*–return", "myreturn"} }
```

Instructing DC to conform to the above rule (BORG) is performed at the command line (or through a script), by using the following command:

```
dc_shell> change_names –hierarchy –rules BORG
```

In addition to the above, users may also desire to alter the bus naming style in the netlist. DC provides a variable through which the user is allowed to tailor the naming style of the busses, written out in the netlist. The variable may again be set in the setup file as follows:

```
bus_naming_style = %s[%d]
```

9.1.3 Remove Unconnected Ports

Many designs suffer from the problem of ports of a block that are left unconnected intentionally, or maybe due to legacy reasons. Although, this practice has no affect on DC in terms of producing functionally correct netlist, however, some designers prefer to remove these ports during synthesis. This is generally a good practice since, if left uncorrected, DC will issue a warning message regarding the unconnected ports. Because a design may contain many such unconnected ports, it is possible that a real warning may get lost between the numerous unconnected ports warnings. It is therefore preferable to remove the unconnected ports and check the design, before generating the netlist. The following commands perform this:

dc_shell> remove_unconnected_ports find(-hierarchy cell, "*")

dc_shell> check_design

9.1.4 Visible Port Names

Generally, all synthesized designs result in mapped components that have one (or more) of their output ports not connected to a net. When DC generates a Verilog netlist, it does not write out the unconnected port names. Depending upon the layout tool, a mismatch might occur between the number of ports in the physical cell versus the number of ports of the same cell present in the netlist. For example, a D flip-flop containing 4 ports namely, D, CLK, Q and QN, may be connected as follows:

DFF dff_reg (.D(data), .CLK(m_clock), .Q(data_out)) ;

In the above case, DC does not write out QN the port, since the function of the inverting QN output is not utilized in the design. Physically, this cell contains all 4 ports, therefore, when the netlist is read in the layout tool, a mismatch between the number of ports occurs. Setting the value of the following variable to true in the setup file can prevent this mismatch:

verilogout_show_unconnected_pins = true

Making the port names visible is solely dependent on the layout tool's requirements. However, recently some layout tool vendors upon realizing this limitation have improved their tools so that the above restriction is not imposed.

9.1.5 Verilog Specific Statements

Some layout tools have difficulty reading the netlist that contains *tri* wires, *tran* primitives and *assign* statements. These are Verilog specific primitives and statements that are generated in the netlist for many possible reasons.

DC generates *tri* wires for designs containing "inout" type ports. For designs containing these types of ports, DC needs to assign values to the bi-directional port, thus producing *tri* wire statement and *tran* primitives. To prevent DC from generating these, users may use the following IO variable in the setup file. When set to true, all tri-state nets are declared as *wire* instead of *tri*.

> verilogout_no_tri = true

Several factors influence the generation of the *assign* statements. Feedthroughs in the design are considered as one such factor. The feedthroughs may occur if the block contains an input port that is directly connected to the output port of the same block. This results in DC generating an *assign* statement in the Verilog netlist. Also the *assign* statements get generated, if an output port is connected to ground, or is being driven by a constant (e.g., 1'b0 or 1'b1). While writing out the netlist in Verilog format, DC issues a warning, stating that the *assign* statements are being written out.

In case of the feedthroughs, the user can prevent DC from generating these statements by inserting a buffer between the previously connected input and output port. This isolates the input port from the output port, thereby breaking the feedthrough. To perform this, the following variable can be used before compiling the design.

> dc_shell> set_fix_multiple_port_nets –feedthroughs

The –buffer_constants option may also be used in the above variable in order to buffer the constants driving the output port. However, since there are many other variations that may produce the *assign* statements, it may be safer to use the following for full coverage:

    dc_shell> set_fix_multiple_port_nets –all –buffer_constants

Many designers complain that *assign* statements get generated in the netlist, even after all the steps described above have been performed. In almost all cases this is caused by the dont_touch attribute present on a net without the users' knowledge. The user can find the presence of this attribute by performing a report_net command. The dont_touch attribute on the net can be removed from the net by using the following command:

    dc_shell> remove_attribute find(net, <net name>) dont_touch

9.1.6 Unintentional Clock or Reset Gating

It is always a good idea to check and double-check the clocks in the design before handing the netlist over for place and route. Remember that the clock provides the reference for all signals i.e., all signals are directly related to the clock and are optimized with respect to it. If the clock is unintentionally buffered (maybe you forgot to apply a set_dont_touch_network attribute on it), it will affect clock latency and skew, which may result in the user not being able to meet the set timing objectives.

Generally, resets are not considered as important as clocks. However, since the set_dont_touch_network attribute is also applied for them, it is wise to check their buffering.

To check for unintentional gating for the clocks, the user may use the following command:

    dc_shell> report_transitive_fanout –clock_tree

To check for unintentional gating of another signal (say, the reset signal), you may use the –from option in the above command. For example:

```
dc_shell> report_transitive_fanout –from reset
```

Obviously, the clocks should be defined before the –clock_tree option can be used. Alternatively, one may also use the –from option for the clocks. This does not require the clocks to be defined first. Note that the –from and the –clock_tree option cannot be used simultaneously.

9.1.7 Unresolved References

Designers should exercise caution and always check for any unresolved references. DC issues a warning for a design containing instantiations of a block that does not have a corresponding definition. For example, block A is the top level module that instantiates sub-block B. If you fail to read the definition of block B in DC while writing out the netlist for block A, DC will generate a warning stating that block A contains unresolved references. Also, this message is issued for cases where a port mismatch occurs between the instanced cell and its definition.

9.2 Layout

With a clean and optimized netlist, the user is ready to transfer the design to its physical form, using the layout tool. Although, layout is a complex process, it can be condensed to three basic steps, as follows:

a) Floorplanning.
b) Clock tree insertion.
c) Routing the database.

9.2.1 Floorplanning

This is considered to be the most critical step within the entire layout process. Primarily a design is floorplanned in order to achieve minimum possible area, while still meeting timing requirements. Also, floorplanning is performed to divide the design into manageable blocks.

In a broad sense, floorplanning consists of placement of cells and macros (e.g., RAMs and ROMs or sub-blocks) in their proper locations. The objective is to reduce net RC delays and routing capacitances, thereby producing faster designs. Placing cells and macros in proper locations also helps produce minimum area and decrease routing congestion.

Almost all designs undergo the floorplanning phase, and time should be spent trying to find the correct placement location of the cells. Optimal placement improves the overall quality of the design. It also helps in reduced synthesis-layout iterations. For small and/or slow designs the floorplanning may not be as important, as that for large and/or timing critical designs consisting of thousands of gates (>150K). For these designs, it is recommended that a hierarchical placement and routing of the design be performed. For example, a sub-block has been placed and routed, meeting all timing and area requirements. The sub-block is subsequently brought in as a fixed macro inside the full design, to be routed with the rest of the cells or macros.

9.2.1.1 Timing Driven Placement

Finding correct locations of cells and macros is time consuming, since each pass requires full timing analysis and verification. If the design fails timing requirements, it is re-floorplanned. This obviously is a time consuming and often frustrating method. To alleviate this, the layout tool vendors have introduced the concept of timing-driven placement, more commonly referred to as timing-driven layout (TDL).

The TDL method consists of forward annotating the timing information of the design generated by DC, to the layout tool. When using this method, the physical placement of cells is dictated by the timing constraints. The layout tools gives priority to timing while placing the cells, and tries not to violate the path constraints.

DC generates the timing constraints in SDF format using the following command:

```
write_constraints  –format <sdf | sdf-v2.1>
                   –cover_design
                   –from <from list>
                   –to <to list>
                   –through <through list>
                   –output <output file name>
```

The above command generates the constraints file in SDF format. Both versions 1.0 and 2.1 are supported. If the layout tool does not support the 2.1 version, then the user may always use the default version 1.0 by specifying "sdf" instead of "sdf-v2.1".

The **write_constraints** command provides many more options in addition to the one illustrated in the above example, however, the use of –cover_design option is more prevalent. The –cover_design option instructs DC to output just enough timing constraints so as to cover the worst path through every driver-load pin pair in the design. For additional information regarding this command and its options, the user is advised to refer to the DC reference manual.

A timing constraint file in SDF version 2.1 format generated by DC with the –cover_design option, is illustrated in Example 9.1. The SDF file contains the TIMINGCHECK field containing PATHCONSTRAINT for all the paths in a design. The last field of the PATHCONSTRAINT timingcheck contains three sets of numbers that define the path delay for a particular path segment. The three numbers, although the same in this example, correspond to minimum, typical, and maximum delay values. These numbers and their corresponding paths govern the placement of cells during layout.

Example 9.1

```
(DELAYFILE
(SDFVERSION "OVI 2.1")
(DESIGN "hello")
(DATE "Mon Jul 20 22:59:49 1998")
(VENDOR "Enterprise")
(PROGRAM "Synopsys Design Compiler cmos")
```

```
(VERSION "1998.02-2")
(DIVIDER /)
(VOLTAGE 2.70:2.70:2.70)
(PROCESS "TYPICAL")
(TEMPERATURE 95.00:95.00:95.00)
(TIMESCALE 1ns)
(CELL
   (CELLTYPE "hello")
   (INSTANCE)
   (TIMINGCHECK
      (PATHCONSTRAINT INPUT1 U751/A3 U751/ZN U754/I1
          U754/ZN REG0/D (1.523:1.523:1.523) )

      (PATHCONSTRAINT INPUT2 U744/A1 U744/Z U745/A1
          U745/ZN REG1/D (1.594:1.594:1.594) )

      (PATHCONSTRAINT REG1/CLK REG1/Q U737/I U737/ZN
          OUTPUT1 ( 3.000:3.000:3.000) )

      (PATHCONSTRAINT REG2/CLK REG2/Q U1131/A2
          U1131/ZN REG3/D (25.523:25.523:25.523) )
                           .
                           .
```

It must be noted that depending upon the size of the design, the generation of the timing constraints for the entire design may take a considerable amount of time. Constraints may be generated for selected timing-critical paths (using –from, –to and –through options) in order to avoid this problem. Alternatively, users may perform hierarchical place and route, where small sub-blocks are routed first, utilizing the TDL method. Hierarchical place and route is a preferred approach, since it is based upon the "divide and conquer" technique. Dividing the chip into small manageable blocks makes it relatively simpler for designers to tackle the run-time problems.

An alternative approach of performing TDL is to let the layout tool generate the timing constraints based upon the boundary conditions, top-level constraints and timing exceptions of the design. This is a tool dependent feature and supported by some layout tool vendors, it may not be supported by others. The layout tool uses its own delay calculator to find out the timing constraints for each path in the design in order to place cells. This method is far superior than the others described previously in the sense that this method is considerably faster, however, a major drawback with this approach is that users are now compelled to use and trust the delay calculator of the layout tool. In any case, timing convergence can be achieved with relative ease, using this approach.

Performing TDL may also have an impact on the overall area. One may find that the area increases when the above approach is used. However, this point is debatable with some users insisting that the total area gets reduced because of the rubber-band effect caused by the TDL method, while others swear by the opposite.

9.2.1.2 Back Annotation of Floorplan Information

Total integration with the back-end tools allows DC to perform with increased efficiency, in order to achieve timing and area convergence. DC makes use of several formats that allow the layout information to be read by DC. For post-layout optimization, it is necessary for DC to know the physical location of each sub-block. Using the physical design exchange format (PDEF) grants DC access to this pertinent information. The PDEF file contains the cluster (physical grouping) information and location of cells in the layout.

Pre-placement, the netlist is optimized using the wire-load models, spread across the logical hierarchy. However, physical hierarchy may be different than the logical hierarchy. Physically, the cells/macros may be grouped depending on the pad locations or some other consideration. Therefore, it is imperative for DC to receive the physical placement information, for it to more effectively optimize the design. This is done by re-adjusting the application of the wire-loads on the design, based upon the physical hierarchy.

DC uses the following command to read the physical placement information generated by the layout tool in PDEF format:

>   read_clusters –design <design name> <pdef filename>

Once the netlist has been re-optimized, the physical information may be passed back to the layout tool through the PDEF file. The following command in DC, performs this task:

>   write_clusters –design <design name> –output <pdef filename>

9.2.1.3 Recommendations

a) In general TDL performs well on all types of designs. However, definitely use TDL for timing critical, and/or high-speed designs, in order to minimize synthesis-layout iterations and achieve timing convergence.

b) When handling large designs, generate timing constraints only for selected nets. This will save you a considerable amount of time. However, if your layout tool is capable of generating its own timing constraints, then it should be given preference over the other approach, in order to save time.

c) Perform hierarchical place and route for large designs. Although tedious, it will generally provide you with best results as well as better control of the overall flow. Hierarchical place and route also expedites hand editing of netlist that is sometimes required after routing is completed.

d) Always use physical placement information in PDEF format while performing post-layout optimization within DC, especially for large hierarchical designs.

9.2.2 Clock Tree Insertion

As explained in previous chapters, it is essential to control the clock latency and skew. Although, some designs may actually take advantage of the positive skew to reduce power, most designs however, require minimal clock skew and clock latency. Larger values of clock skew cause race conditions that increase the chance of wrong data being clocked in the flops. Controlling the skew and latency requires a lot of effort and foresight.

As mentioned before, the layout tool performs the clock tree synthesis (CTS for short). The CTS is performed immediately after the placement of the cells, and before routing these cells. With input from the designer, the layout tool determines the best placement and style of the clock tree. Generally, designers are asked for the number of levels along with the types of buffers used for each level of the clock tree. Obviously, the number of levels is dependent on the fanout of the clock signal.

In a broad sense, the number of levels of the clock tree is inversely proportional to the drive strength of the gates used in the clock tree. In other words, you will need more levels, if low drive strength gates are used, while the number of levels is reduced if high drive strength gates are used.

To minimize the clock skew and clock latency, designers may find the following recommendations helpful. It must be noted that these recommendations are not hard and fast rules. Designers often resort to using a mixture of techniques to solve the clocking issues.

a) Use a balanced clock tree structure with minimum number of levels possible. Try not to go overboard with the number of levels. The more the levels, the greater the clock latency.

b) Use high drive strength buffers in large clock trees. This also helps in reducing the number of levels.

c) In order to reduce clock skew between different clock domains, try balancing the number of levels and types of gates used in each clock tree. For instance, if one clock is driving 50 flops while the other clock is driving 500 flops, then use low drive strength gates in the clock tree of the first clock, and high drive strength gates for the other. The idea here is to speed-up the clock driving 500 flops, and slow down the clock that is driving 50 flops, in order to match the delay between the two clock trees.

d) If your library contains balanced rise and fall buffers, you may prefer to use these instead. Remember, in general it is not always true that the balanced rise and fall buffers, are faster (less cell delay) than the normal buffers. Some libraries provide buffers that have lower cell delays for rise times of signals, as compared to the fall times. For designs utilizing the positive edge trigger flops, these buffers may be an ideal choice. The idea is to study the library and choose the most appropriate gate available. Past experience also comes in handy.

e) To reduce clock latency, you may try to use high drive inverters for two levels. This is because, logically a single buffer cell consists of two inverters connected together, and therefore has an cell delay of two inverters. Using two separate inverters (two levels) will achieve the same function, but will result in reduced overall cell delay – since you are not using another buffer (2 more inverters) for the second level. Use this approach, only for designs that do not contain gated clocks. The reason for this explained later (point h).

f) Do not restrict yourself to using the same type and drive strength gate for CTS. Current layout tools allow you to mix and match.

g) For a balanced clock tree (e.g., 3 levels), the first level is generally a single buffer driven by the Pad. In order to reduce clock skew, the first level buffer is placed near the center of the chip, so that it can connect to the next level of buffers, through equal interconnect wires. This creates a ring like structure with the first buffer in the center, with the second set of buffers (second level) surrounding it, and the last stage surrounding the second level. Thus, the distance between the first, second and the third level are kept at minimum. However, although a good arrangement, it does result in the first level buffer being placed farthest from the source (Pad). If a minimum size wire is used to route the clock network from the Pad source to the first buffer, it will result in a large RC delay that will affect the clock latency. Therefore, it is necessary to size-up (widen) this wire from the Pad source to the input of the buffer (first level), in order to reduce the resistance of the wire, thereby reducing the overall latency. Depending upon the size of your design and the number of levels, you may also need to perform this operation on other levels.

h) In order to minimize the skew, the layout tool should have the ability to tap the clock signal, from any level of the clock tree. This is especially

LINKS TO LAYOUT AND POST LAYOUT OPTIMIZATION   175

important for designs that contain gated clocks. If the same clock is used for other ungated flops, then it results in additional delay, hence the skew. If the clock tree ended at the gate, the additional delay will cause a large skew between the gated-clock flop and the ungated-clock flop as shown in Figure 9-1(a). Therefore it is necessary to tap the clock source from a level up for the gated-clock flop, while maintaining the full clock tree for the ungated clock flop, as illustrated in Figure 9-1 (b). However, if inverters are used in the clock tree (point e), then the above approach breaks down. In this case, do not use inverters as part of the clock tree.

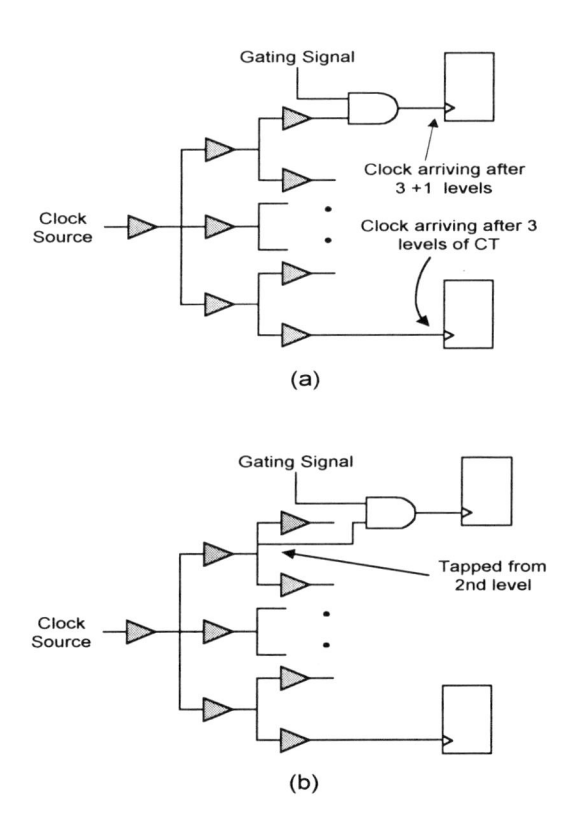

Figure 9-1. CTS for Gated Clocks versus Non-gated Clocks

9.2.3 Transfer of Clock Tree to Design Compiler

Clock tree synthesis done by the layout tool modifies the physical design (cells are added in the clock network). This modification is absent from the original netlist present in DC. Therefore, it is necessary for the user to accurately transfer this information to DC. There are several ways to do this.

a) Generally all layout tools have the capability to write out the design in EDIF or Verilog format. Since, everything may appear flat to the layout tool, designers may receive a flat netlist from the layout tool. Of course, this netlist will contain the clock tree information, but the enormous size of the netlist itself may be daunting and unmanageable. Furthermore, due to the absence of original design hierarchy, the flat netlist is not easily readable. Another problem with this approach is that the user is now forced to designate this netlist as the "golden" netlist, meaning that all verification (LVS etc.) has to be performed against this netlist. Doing this is comparable to "digging your own grave" because, if the layout tool botches the layout, the same anomalies will be reflected in the netlist. Of course, the LVS will pass without flagging any errors, since the user is checking physical layout data against layout generated netlist i.e., performing LVL instead of LVS. An alternative is to perform formal verification between the original hierarchical netlist and the layout generated flat netlist. This certainly is a viable approach, but has its own limitations regarding the size and complexity of the design. Most formal verification tools suffer from this limitation, i.e., they excel in individual block verification, but fall short in full chip verification. This is especially true for verifying flat netlists against hierarchical netlists.

b) The second approach is to only transfer, point-to-point clock delay information, starting from the clock source to its endpoints (clock pins of the flops). The delay calculator of the layout tool will perform this task, and upon instruction will provide the designer, the point-to-point timing information of the clock tree in SDF format. Designers may back annotate this SDF file to the original design in order to determine the clock latency and skew. This method does not require the clock tree to be transferred to DC from the layout tool. However, this approach has its own pitfalls. Primarily, this approach does not allow the usage of SPEF data from the layout for back annotation to PT. Furthermore, the designer is now compelled to trust the delay calculator of the layout tool i.e., another variable has been introduced that requires qualification. Since, the layout libraries are separate from Synopsys libraries, in order to get the same delay numbers, the timing numbers present in the Synopsys library need to match exactly to that of the layout library. The dilemma of verifying the original netlist against the layout database still exists, especially since the original netlist does not contain the extra cells and nets due to clock tree insertion. However, one can certainly find work-arounds and may use this approach successfully.

c) A solution to all of the above problems is to creatively transfer the entire clock tree to DC without changing the hierarchy of the design. Some layout tools may even generate Synopsys scripts that contain dc_shell commands like, disconnect_net, create_cell, create_port and connect_net. These commands on execution insert the clock tree into the original design database in DC, while still maintaining the hierarchy. Of course, one needs to verify the resulting modified netlist against the original netlist by performing formal verification. Since the design hierarchy is not altered, the formal verification runs smoothly.

d) Another solution involves brute force modification. Generally the layout tools, upon completion of CTS, produce a summary report of all changes made to the design. One may take advantage of this report and parse it to retrieve the relevant information (e.g., name of clock tree insertion points, type and name of buffers etc.) using scripting languages like Perl or Awk. Once the information is gathered, the original Verilog netlist may be directly modified without going through DC. The modified netlist should be read back into DC to check for any syntax errors. In addition, the modified netlist should also be formally verified against the original netlist.

Recently, upon realizing this problem the layout tool vendors have facilitated this process by generating the hierarchical netlist from the layout database. This netlist contains the clock tree information and should be verified formally against the original netlist. Upon successful verification, the netlist may be declared as "golden".

9.2.4 Routing

After the clock tree insertion, the final step involves routing the chip. In a broad sense, the routing is divided into two phases:

1. Global routing, and
2. Detailed routing.

The first routing phase is called the global route, in which the global router assigns a general pathway through the layout for each net. During global route, the layout surface is divided into several regions. The global router decides the shortest route through each region in the layout surface, without laying the geometric wires.

The second routing phase is called the detailed route. The detailed router makes use of the information gathered by the global route and routes the geometric wires within each region of the layout surface.

It must be noted that if the run-time of global route is long (more than the placement run-time), it indicates a bad placement quality. In this case, the placement should be performed again with emphasis on reduced congestion.

9.2.5 Extraction

Until now, synthesis and optimization was performed utilizing the wire-load models. The wire-load models are based on statistically estimating the final routing capacitances. Because of the statistical nature of wire-load models, they may be completely inaccurate compared to the real delay values of the routed design. This variation between the wire-load models and the real delay values results in an non-optimized design.

The layout database is extracted to produce the delay values necessary to further optimize the design. These values are back annotated to PT for static timing analysis, and to DC for further optimization and refinement of the design.

9.2.5.1 What to Extract?

In general, almost all layout tools are capable of extracting the layout database using various algorithms. These algorithms define the granularity and the accuracy of the extracted values. Depending upon the chosen algorithm and the desired accuracy, the following types of information may be extracted:

1. Detailed parasitics in DSPF or SPEF format.
2. Reduced parasitics in RSPF or SPEF format.
3. Net and cell delays in SDF format.
4. Net delay in SDF format + lumped parasitic capacitances.

The DSPF (Detailed Standard Parasitic Format) contains RC information of each segment (multiple R's and C's) of the routed netlist. This is the most accurate form of extraction. However, due to long extraction times on a full design, this method is not practical. This type of extraction is usually limited to critical nets and clock trees of the design.

The RSPF (Reduced Standard Parasitic Format) represents RC delays in terms of a pi model (2 C's and 1 R). The accuracy of this model is less than that of DSPF, since it does not account for multiple R's and C's associated with each segment of the net. Again, the extraction time may be significant, thus limiting the usage of this type of information. Target applications are critical nets and small blocks of the design.

Both detailed and reduced parasitics can be represented by OVI's (Open Verilog International) Standard Parasitic Exchange Format (SPEF).

The last two (number 3 and 4) are the most common types of extraction used by the designers. Both utilize the SDF format. However, there is major difference between the two. Number 3 uses the SDF to represent both the cell and net delays, whereas number 4 uses the SDF to represent only the net delays. The lumped parasitic capacitances are generated separately. Some layout tools generate the lumped parasitic capacitances in the Synopsys set_load format, thus facilitating direct back annotation to DC or PT.

It is worth mentioning that PT can read all five formats (DSPF, RSPF, SPEF, SDF and set_load), whereas, DC can only read the SDF and set_load file formats. The SDF and set_load file formats are not as accurate as the DSPF or RSPF types of extraction, however, the time to extract the layout database is significantly reduced. For most designs this type of extraction provides sufficient accuracy and precision. However, as suggested, only critical nets and clocks in the design should be targeted for DSPF or RSPF types of extraction.

For the layout tool to generate a full SDF (number 3 approach), it uses its own delay calculator to compute the cell delays that are based upon the output loading and the transition time of the input signal. However, there is a flaw in using this approach. The synthesis was done using DC that used its own delay calculator to optimize the design. By choosing to use the full SDF generated by the layout tool, we are now introducing another variable that needs qualification. How do we know that the delay calculator used by the layout tool is more accurate than the one used by DC? Also, upon back annotation of the full SDF to PT, the full capability of PT is also not being utilized. This is because the cell delays are already fixed in the SDF file, and performing case analysis in PT will not yield accurate results, even if the conditional delays are present in the SDF file. This is discussed at length in Chapter 10.

Another problem exists with the above approach. Since only the cell and net delays are back annotated, DC does not know the parasitic capacitances associated with each net of the design. Therefore, when performing post-layout optimization, DC can only make use of the wire-load models to make incremental changes to the design, thus defeating the whole purpose of back annotation. However, if the fourth approach was used (net delays in SDF format + lumped parasitic capacitances), DC makes use of the net loading information during post-layout optimization (e.g., to size up/down gates).

To avoid these problems, it is recommended that only the <u>net RC delays (also called as interconnect wiring delays) and lumped parasitic capacitances</u> are extracted from the layout database. Upon back annotation, DC or PT uses its own delay calculator to compute the cell delays, based upon the back annotated interconnect RC's and capacitive net loading.

To summarize, it is recommended that the following types of information should be generated from the layout tool for back annotation to DC in order to perform post layout optimization:

a) Net RC delays in SDF format.
b) Capacitive net loading values in **set_load** format.

For static timing analysis, using PT, the following types of information can be generated:

a) Net RC delays in SDF format.
b) Capacitive net loading values in **set_load** format.
c) Parasitic information for clock and other critical nets in DSPF, RSPF or SPEF file formats.

9.2.5.2 Estimated Parasitic Extraction

The extraction of parasitics at the pre-route level (after global routing) provides a closer approximation to the parasitic values of the final routed design. If the estimates indicate a timing problem, it is fairly easy to quickly re-floorplan the design before starting the detailed route. This method reduces synthesis-layout iterations and avoids wastage of valuable time.

The difference between the estimated extracted delay values after the global routing and the real delay values after the detailed routing is minimal. In contrast, the estimated delay values between the floorplan extraction and detailed route extraction may be significant. Therefore it is prudent that after floorplanning, cell placement and clock tree insertion, the design be globally routed, before extracting the estimated delay numbers.

A complete extraction flow is shown in Figure 9-2. If major timing violations exist after global route, it may be necessary to re-optimize the design within DC with estimated delays back annotated. However, if the timing violations are not severe then re-floorplanning (and/or re-placement of cells) the design may achieve the desired result. The detailed routing should be performed, only after the eliminating all timing violations produced after the global routing phase.

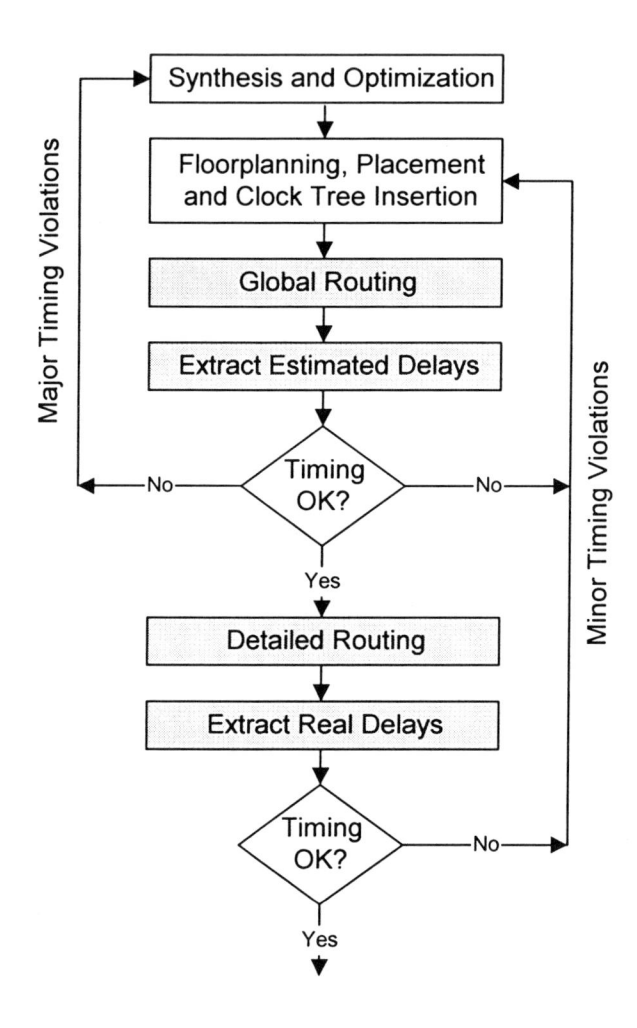

Figure 9-2. Routing and Extraction Flow

9.2.5.3 Real Parasitic Extraction

A full extraction (actual values – no estimation) is performed after the design has been satisfactorily globally routed, i.e., no DRC violations and achievement of required die-size.

This by far is the most critical part of the entire process. The final product may not work, if the extracted values are not accurate. With technologies shrinking to 0.18 micron and below, the extraction algorithm of the layout tool need to take into account the second and third order parasitic effects. Any slight deviations of these values may cause the design to fail.

Consider a case where the extracted values are too pessimistic. Static timing analysis indicates that the signals are meeting hold-time requirements. However, in reality, the signals are arriving faster, causing real hold-time violations, but due to pessimistic back annotated parasitic capacitances, the design is passing static timing. The case for setup-time is similar, if the extracted values are too optimistic.

9.3 Post-Layout Optimization

Post-layout optimization is performed to further optimize and refine the design. The process involves back annotating the data generated by the layout tool, to the design residing in DC. Depending upon the severity of violations, the optimizations may include full synthesis or minor adjustments through the use of in-place optimization (IPO) technique. As explained in the previous section, the layout related data suitable for back annotation to DC are:

a) Net RC delays in SDF format.
b) The **set_load** file, containing capacitive net loading.
c) Physical placement information in PDEF format.

9.3.1 Back Annotation and Custom Wire Loads

The next step involves analyzing the static timing of the design. Designers may choose to perform this step using PT or DC's internal static timing analysis engine. In any case, post layout optimization can only be performed within DC therefore the layout data needs to be back annotated to both DC and PT.

Depending on the process technology, the layout tool may generate two separate files that correspond to the worst and the best case. If there are two separate Synopsys libraries pertaining to each case, then back annotate the worst case layout data to the design using the worst case Synopsys library. Similarly, best case layout data should be back annotated to the design mapped to best case Synopsys library.

Some vendors provide only one Synopsys library that covers all cases, i.e., the library is characterized for TYPICAL case, with the WORST and the BEST case values derived (derated) from the TYPICAL case. In a situation like this, it is recommended that the designer back annotate the worst case numbers to the design with operating conditions set to WORST, in order to perform the worst case timing analysis. The best case timing analysis should be performed with best case timing numbers back annotated to the design with operating conditions set to BEST.

Use the following dc_shell commands to back annotate layout-generated information to the design present in DC, before performing post-layout optimization.

    dc_shell> current_design   <design name>

    dc_shell> include –quiet   <set_load file name>

    dc_shell> read_timing –f sdf   <RC file name in SDF format>

    dc_shell> read_clusters   <cluster file name in PDEF format>

Use the following pt_shell commands to back annotate layout-generated information to the design in PT, before performing static timing analysis.

LINKS TO LAYOUT AND POST LAYOUT OPTIMIZATION

pt_shell> current_design   <design name>

pt_shell> source   <set_load file name in PT format>

pt_shell> read_sdf   <RC file name in SDF format>

pt_shell> read_parasitics   <DSPF, RSPF or SPEF file name>

After back annotation in PT, if the design fails static timing with substantial amount of violations, the user may need to perform re-synthesis (or even re-code certain blocks). Therefore, it is prudent to use the existing layout information for re-synthesis. Discarding the layout data during re-synthesis only wastes the time and effort spent for layout. Furthermore, the layout data is helpful in fine tuning the design. To achieve the maximum benefit, custom wire-load models should be generated from DC, using the existing layout information. The resulting gate level netlist using the custom wire-load models provide a closer match to the post-layout timing results. Use the following dc_shell command to create custom wire-load models:

create_wire_load –design <design name>
                 –cluster <cluster name>
                 –trim <trim value>
                 –percentile <percentile value>
                 –output <output file name>

Although, there are other options available for the above command, generally the ones listed above suffice for most designs. The trim value is used to discard data that falls below a certain value, while the percentile value is used to calculate the average value. By altering the percentile value, one may add optimism or pessimism in the custom wire-load models. The cluster name is obviously the grouping name that was used during layout, to group cells or blocks together.

After the creation of the custom wire-load models (CWLM), the library should be updated to account for the new CWLMs. This is because the original technology library contains only the generic wire-load models that are not particular to a specific design. To use the CWLMs that were generated by the above command, the library must be updated. The following command may be used to update the library present in DC memory:

> update_lib <library name> <CWLM file name>

It must be noted that the above command does not alter or overwrite the source library. It only updates the DC memory to include the new CWLMs.

9.3.2 In-Place Optimization

For designs with minor timing violations after layout, there is no need to perform a full chip synthesis. In-place optimization or IPO is an excellent method to fine-tune a design, in order to eliminate these violations. The concept of IPO is to keep the structure of the design intact while modifying only the failing parts of the design, thereby having a minimal impact on the existing layout. IPO is commonly used to add/swap gates at specific locations to fix setup and/or hold-time problems.

The IPO is library dependent and can be limited to perform only the following:

a) Resize cells.
b) Insert or delete existing cells (mainly buffers).

Usually, all Synopsys technology libraries have an attribute defined that enables or disables the IPO. The attribute and its value, that enables IPO in a library is:

> in_place_swap_mode : match_footprint

Along with the above library level attribute, all cells in the Synopsys library also have the cell_footprint information. For example, two cells with same functionality, but different drive strengths may have the same cell_footprint value. This means that the two cells have identical physical coverage area, therefore replacing one for the other will not impact the existing layout, i.e., adjacent cells will not shift. However, this restriction, along with cell sizing, area optimization and buffer insertion, is controlled by the use of the following variables:

```
compile_ignore_footprint_during_inplace_opt    = true | false
compile_ok_to_buffer_during_inplace_opt        = true | false
compile_ignore_area_during_inplace_opt         = true | false
compile_disable_area_opt_during_inplace_opt    = true | false
```

Using these variables allows the designer the ability control the amount of changes made in the design. The appropriate values of the above variables may be set before performing IPO at dc_shell command line; or in the Synopsys setup file.

The IPO is invoked by using the following commands:

dc_shell> compile –in_place

dc_shell> reoptimize_design –in_place

Both commands are similar in nearly all respects, i.e., both make use of the back annotated layout information, with the exception of physical information. The reoptimize_design makes use of the physical location information while re-optimizing the design. Another difference between the two commands is that the "compile –in_place" command uses the library wire-load models during IPO, whereas the "reoptimize_design –in_place" command makes use of the custom wire-load models. Therefore, it is imperative, that the latter command be used while performing IPO.

It must be noted that the reoptimize_design when used on its own makes major modifications to the design. To eliminate this possibility, always use the –in_place option.

9.3.3   Location Based Optimization

Location Based Optimization or LBO is an integral part of IPO, and is invoked automatically while performing IPO for designs containing back annotated physical placement location information in PDEF format. In this chapter the LBO is being identified separately from IPO due to its importance and additional capability.

Performing IPO with LBO improves the overall optimization of the design, since DC now has access to the cell placement information. This allows DC to apply more powerful algorithms during optimization.

Consider a path segment starting from primary input and ending at a flop. Post-layout timing analysis reveals a hold-time problem for this path. As shown in Figure 9-3, LBO optimization will add buffer(s) near the flop (endpoint), instead of adding it at the source (startpoint). Adding buffer(s) at the source may cause a setup-time failure for another path originating from the same source.

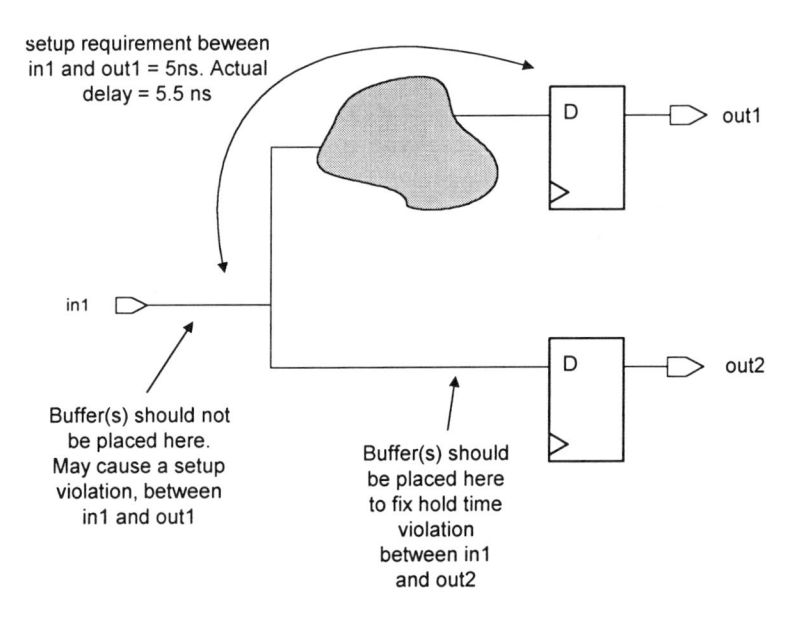

Figure 9-3. IPO versus LBO

In addition to the enhanced capability of inserting buffers at optimal locations, LBO also provides better modeling of cross cluster nets, and the new nets that were created to connect the inserted or deleted buffers. This is due to the fact that, DC is aware of the location where the cells were inserted or deleted.

In order for the "reoptimize_design –in_place" to perform LBO, the following variables need to be set to true; in addition to the IPO related variable "compile_ok_to_buffer_during_inplace_opt":

>lbo_buffer_removal_enabled = true

>lbo_buffer_insertion_enabled = true

LBO is not enabled for buffer insertion or removal, if the value of above variables is set to false (default). Location information is disregarded for this case, with only IPO algorithms used to handle the buffer insertion or deletion.

The changes made by performing IPO and/or LBO, using both "compile –in_place" or "reoptimize_design –in_place", can be written out to a file by using the following dc_shell variable:

>reoptimize_design_changed_list_file_name = <file name>

If the file already exists, the new set of changes will be appended to the same file.

9.3.4 Fixing Hold-Time Violations

Nearly every design undergoes the process of fixing hold-time violations, especially for faster technologies. Designers tackle this problem using various approaches. For this reason, a separate section is devoted to discuss issues arising from using these methods, and to consolidate them under one topic.

Most designers synthesize the design with tight constraints in order to maximize the setup-time. The resulting effect is a fast logic with data arriving faster at the input of the flop, with respect to the clock. This may result in hold-time violations due to data changing value before being latched by the flop. Generally, designers prefer to fix the hold-time violations after initial placement and routing of the design, thereby making use of more accurate delay numbers.

Removing hold-time violations involves delaying the data with respect to the clock, so that the data does not change for a specified amount of time (hold-time) after the arrival of the clock edge. There are several methods utilized by designers to insert the appropriate delays, as outlined below:

a) Using Synopsys methodology.
b) Inserting delays manually.
c) Inserting delays automatically, by using brute force dc_shell commands.

9.3.4.1 Synopsys Methodology

Synopsys provides the following dc_shell command, which enables the compile command to fix the hold-time violations:

    set_fix_hold <clock name>

    dc_shell> set_fix_hold CLK

The above command may be used during initial compile (or post-layout), by setting the min/max library concurrently (version DC98 onwards), and specifying the min/max values for set_input_delay command. The idea behind setting the min/max library at the same time is to eliminate the two-pass (initial synthesis for maximum setup-time and re-optimization to fix hold-time violations) synthesis needed for almost all designs. Example 9.2 illustrates the methodology of fixing the post-layout hold-time violations using the single pass synthesis approach.

Example 9.2

```
set_min_library  "<worst case library name>"         \
          –min_version "<best case library name>"

set_operating_conditions –min BEST –max WORST

include net_delay.set_load
read_timing interconnect.sdf
read_clusters floorplan.pdef

set_input_delay –max 20.0 –clock CLK {IN1 IN2}
set_input_delay –min –1.0 –clock CLK {IN1 IN2}

set_output_delay –max 10 –clock CLK all_outputs()

set_fix_hold CLK

reoptimize_design –in_place
```

Alternatively, the design may be compiled with maximum setup-time using the worst case library, followed by re-optimization after layout, in order to fix the hold-time violations by mapping the design to the best case library. Although, this method uses the two-pass synthesis approach, it is still recommended because of its stable nature. Most designers prefer to use this approach because of the time and effort that has been invested in defining and maturing this methodology.

It must be noted that the above command is independent of IPO commands. The IPO commands are generally used to fix hold-time violations after initial layout with layout information back annotated to the design. Fixing pre-layout hold-time violations is accomplished by compiling the design incrementally, using the "compile –incremental" command.

The **set_fix_hold** command instructs DC to fix the hold-time violations by inserting the buffers at appropriate locations. This again is controlled by the IPO related variables described previously. With buffer insertion disabled, the cells in the data path can only be swapped (i.e., replacing a higher drive strength gate with a lower drive strength gate) to increase the cell delay, thereby delaying the data arriving at the flop input.

9.3.4.2 Manual Insertion of Delays

If the timing analysis reveals a very small number of hold-time violations (less than 10 to 20 places), it may not be worthwhile to fix these violations using the set_fix_hold command. The delays in this case may be manually inserted in the netlist. The designer may chain a string of buffers to delay the data, with respect to the clock just enough, so that it passes the hold-time checks.

A point to note however is that a chain of buffers may not be able to provide adequate delay, since the delay is dependent on the placement of the buffers in the layout. Generally, the buffers will be placed very close to each other therefore the total delay will be the sum of the delay through each cell. The interconnect delay itself will be insignificant, due to the close proximity of the placed cells. To overcome this, it is recommended to link a number of high-fanin gates (e.g., 8 input AND gate) with all inputs tied together, and connected to form a chain. The advantage of using this approach is that now the input pin capacitance of the high fanin gates is being utilized. The input pin capacitance of a high fanin gate is usually much larger than a single input buffer. Therefore, this method of delaying the data provides a solution that is independent of the cell placement location in the layout. This approach is suitable if the technology library does not contain delay cells. If these cells are present, then they should be targeted to fix the hold-time violations.

9.3.4.3 Brute Force Method

This is a unique method, but requires expertise in scripting languages like Perl or Awk.

For instance, if the timing report shows many hold-time violations, and fixing them through Synopsys methodology means large run times. In this case, an alternative approach, to find the amount of slack (setup-time analysis) and the corresponding violation (hold-time analysis) of the failing paths, is to parse the timing report (both for worst-case and best case) using a scripting language. Using these numbers, the user may generate dc_shell commands like, disconnect_net, create_cell and connect_net, for the failing paths. Upon execution of these commands in dc_shell, it will force DC to insert and connect buffers at appropriate places (should be done near the endpoints of the failing paths). This is called a brute force method, done automatically.

This by no means is a clean approach, but works remarkably well. The time taken to fix hold-time violations using this approach is negligible as compared to the Synopsys methodology.

9.4 Future Directions

The future holds considerable challenges to the EDA industry. The problem is compounded by the fact that layout tool vendors lack synthesis technology while Synopsys lacks the back-end expertise. Nearly all back-end tool vendors provide interface to Synopsys (through LTL), but total integration is missing.

Time-to-market is rapidly shrinking while design complexities are increasing. The problem is further aggravated by shrinking geometries, forcing ASIC designers to think about power and cross-talk along with timing, much earlier in the design cycle. The exchange of data between layout tools and DC (and PT) is certainly not efficient. Time wasted during synthesis-layout iterations is still a major bottleneck.

In future, especially for VDSM technologies, the back-end tool vendors will be left with no other choice but to incorporate the synthesis technology. Many loyal Synopsys users will still prefer to perform synthesis and pre-layout optimizations of their designs using Synopsys tools, however, the post-layout optimization may move into the layout tools arena, provided they gain the synthesis know-how.

Post-layout optimization frequently involves insertion and/or deletion of cells in order to fix the timing problems. Therefore, the layout tools have a built-in advantage over Synopsys, as they do not need the PDEF information to find the placement location of each cell, thus the exchange of data back and forth between the layout tools and DC is eliminated. Furthermore, almost all layout tools incorporate their own delay calculator, which further simplifies the process of swapping, adding, or removing of the cells, in order to fix timing problems.

Synopsys, upon realizing these problems is currently in the process of introducing a completed flow to layout, termed as physical synthesis, which includes tight integration of synthesis with the place and route engine. This may provide a major change in the current methodology. Most designers may prefer this approach in order to maintain the golden nature of their netlist, rather than letting the layout tools handle the post-layout optimization. Only time will tell.

9.5 Chapter Summary

Links to layout is an important part of the integration between the layout tool and DC. This chapter focussed on all aspects of exchanging data to and from layout tools, in order for DC to perform better optimization and fine-tuning the design.

Issues related to transfer of clock tree information from the layout tool to DC were explained in detail. Cross checking the netlist generated by the layout tool against the original netlist remains a major bottleneck. Various alternatives were provided to the user in order to overcome this issue and choose the right solution.

Starting from how to generate a clean netlist from DC in order to minimize layout problems, this chapter covered placement and floorplanning, clock tree insertion, routing, extraction, and post-layout optimization techniques, including various methods to fix the hold-time violations. At each step, recommendations are provided to facilitate the user in choosing the right direction.

As we move towards VDSM technologies, a future trend towards total integration between synthesis and back-end tools is also discussed.

10

SDF GENERATION
For Dynamic Timing Simulation

The standard delay format or SDF contains timing information of all the cells in the design. It is used to provide timing information for simulating the gate-level netlist.

As mentioned in Chapter 1, verification of the gate-level netlist of a design through dynamic simulation is not a recommended approach. Dynamic simulation method is used to verify the functionality of the design at the RTL level only. Verification of a gate-level design using dynamic simulation depends solely on the coverage provided by the test-bench, therefore, certain paths in the design that are not sensitized will not get tested. In contrast, formal verification techniques provide superior validation of the design.

The dynamic simulation method for gate-level design verification is still dominant, and is used extensively by designers. Due to this reason, this chapter provides a brief description on generating the SDF file from DC and PT, which can be used to perform dynamic timing simulation of the design. Furthermore, a few innovative ideas and suggestions are provided to facilitate designers in performing successful simulation.

Please note that some of the information provided in this chapter is also described in previous chapters. Since, this is an important topic, it is deemed necessary that a full chapter relating to SDF generation be devoted for the sake of clarity and completeness.

10.1 SDF File

The SDF file contains timing information of each cell in the design. The basic timing data comprises of the following:

a) IOPATH delay.
b) INTERCONNECT delay.
c) SETUP timing check.
d) HOLD timing check.

Following, is an example SDF file that contains the timing information for two cells (sequential cell feeding an AND gate), along with the interconnect delay between them:

```
(DELAYFILE
(SDFVERSION "OVI 2.1")
(DESIGN "top_level")
(DATE "Dec 30 1997")
(VENDOR "std_cell_lib")
(PROGRAM "Synopsys Design Compiler cmos")
(VERSION "1998.08")
(DIVIDER /)
(VOLTAGE 2.70:2.70:2.70)
(PROCESS "WORST")
(TEMPERATURE 100.00:100.00:100.00)
(TIMESCALE 1ns)
(CELL
   (CELLTYPE "top_level")
   (INSTANCE)
   (DELAY
     (ABSOLUTE
```

```
      (INTERCONNECT sub1/U1/Q sub1/U2/A1 (0.02:0.03:0.04)
                                         (0.03:0.04:0.05))
       )
     )
   )
   (CELL
     (CELLTYPE "dff1")
     (INSTANCE sub1/U1)
     (DELAY
       (ABSOLUTE
       (IOPATH CLK Q (0.1:0.2:0.3) (0.1:0.2:0.3))
       )
     )
     (TIMINGCHECK
       (SETUP (posedge D) (posedge CLK) (0.5:0.5:0.5))
       (SETUP (negedge D) (posedge CLK) (0.6:0.6:0.6))
       (HOLD (posedge D) (posedge CLK) (0.001:0.001:0.001))
       (HOLD (negedge D) (posedge CLK) (0.001:0.001:0.001))
     )
   )
   (CELL
     (CELLTYPE "and2")
     (INSTANCE sub1/U2)
     (DELAY
       (ABSOLUTE
       (IOPATH A1 Z (0.16:0.24:0.34) (0.12:0.23:0.32))
       (IOPATH A2 Z (0.11:0.21:0.32) (0.17:0.22:0.34))
       )
     )
   )
 )
)
```

The IOPATH delay specifies the cell delay. Its computation is based upon the output wire loading and the transition time of the input signal.

The INTERCONNECT delay is a path based, point-to-point delay, which accounts for the RC delay between the driving gate and the driven gate. This wire delay is specified from the output pin of the driving cell to the input pin of the driven cell.

The SETUP and HOLD timing checks contain values that determine the required setup and hold-time of each sequential cell. These numbers are based upon the characterized values in the technology library.

10.2 SDF File Generation

The SDF file may be generated for pre-layout or post-layout simulations. The post-layout SDF is generated from DC or PT, after back annotating the extracted RC delay values and parasitic capacitances, to DC or PT. The post-layout values thus represent the actual delays associated with the design. The following commands may be used to generate the SDF file:

DC Command

write_timing –format sdf-v2.1 –output <filename>

PT Command

write_sdf –version [1.0 or 2.1] <filename>

Note: By default, PT generates the 2.1 version of SDF.

10.2.1 Generating Pre-Layout SDF File

The pre-layout numbers contain delay values that are based upon the wire-load models. Also, the pre-layout netlist does not contain the clock tree. Therefore, it is necessary to approximate the post-route clock tree delays while generating the pre-layout SDF.

In order to generate the pre-layout SDF, the following commands approximate the post-route clock tree values by defining the clock delay, skew, and the transition time.

DC Commands

create_clock –period 30 –waveform {0 15} CLK

set_clock_skew –delay 2.0 CLK

set_clock_transition 0.2 CLK

PT Commands

create_clock –period 30 –waveform [list 0 15] [list CLK]

set_clock_latency 2.0 [get_clocks CLK]

set_clock_transition 0.2 [get_clocks CLK]

By setting (fixing) these values as illustrated above, designers may assume that the resulting SDF file will also contain these values, i.e., the clock delay from the source to the endpoint (clock input port of the flops) is fixed at 2.0. However, this is not the case. DC only uses the above commands to perform static timing analysis, and does not output this information to the SDF file. To avoid this problem, designer should force DC to use the specified delay value instead of calculating its own. To ensure the inclusion of 2.0ns as the clock delay, a dc_shell command (explained later in the section) is used to "massage" the resulting SDF.

At the pre-layout level, the clock transition time should also be specified. Failure to fix the transition time of the clock results in false values computed for the driven flops. Again, the culprit is the missing clock tree at the pre-layout level. The absence of clock tree forces the assumption of high fanout for source clock, which in turn causes DC to compute slow transition times for the entire clock network. The slow transition times will affect the driven flops (endpoint flops), resulting in large delay values computed for them.

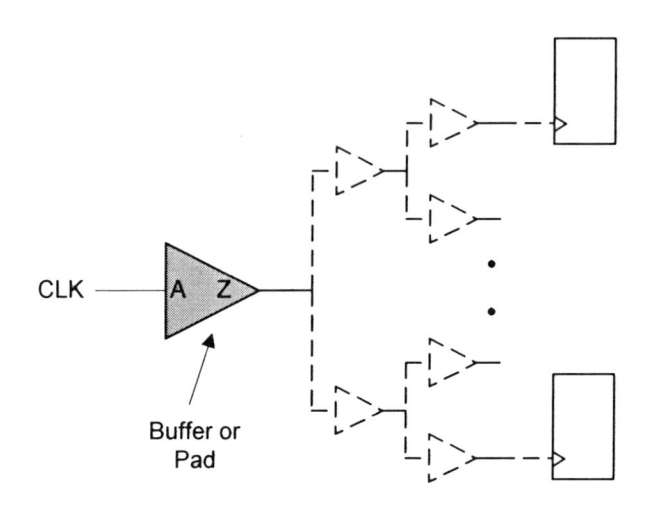

Figure 10-1. Specifying Clock Tree Delay

Consider the diagram shown in Figure 10-1. The dotted lines illustrate the placement of clock tree buffers, synthesized during layout. At pre-layout level, these buffers do not yet exist. However, there usually is a buffer/cell (shown as shaded cell) at the clock source. This cell may be a big driver, instantiated by the designer with the sole purpose of driving the future clock tree, or it may simply be an input pad. Let us assume that this is an input pad (called CLKPAD) with input pin A and output pin Z. At the pre-layout level, the output pin Z connects directly to all the endpoints.

The easiest way to fix the SDF, so that it reflects the 2.0 ns clock delay from the source "CLK" to all the endpoints, is to replace the delay value of the shaded cell (from pin A to pin Z) calculated by DC, with 2.0 ns. This can be achieved by using the following dc_shell command:

```
dc_shell> set_annotated_delay 2.0 –cell      \
                –from CLKPAD/A –to CLKPAD/Z
```

☺ Note: A similar command also exists for PT.

SDF GENERATION

The above command replaces the value calculated by DC, with the one specified i.e., 2.0 ns. This delay gets reflected in the SDF file in the form of IOPATH delay, for the cell CLKPAD, from pin A to pin Z.

Fixing the delay value of the clock solves the problem of clock latency. However, what happens to the delay values of the driven flops? Designers may incorrectly assume that DC uses the specified clock transition for the sole purpose of performing static timing analysis, and may not use the specified values to calculate delays of the driven flops. This is not so. DC uses the fixed transition value of the clock to calculate delays of driven gates. Not only are the transition values used to perform static timing analysis, but they are also used while computing the delays of the driven cells. Thus the SDF file contains the delay values that are approximated by the designer at the pre-layout phase.

10.2.2 Generating Post-Layout SDF File

The post-layout design contains the clock tree information. Therefore, all the steps that were needed to fix the clock latency, skew, and clock transition time, during pre-layout phase, are not required for post-layout SDF file generation. Instead, the clock is propagated through the clock network to provide the real delays and transition times.

As explained in Chapter 9, only the extracted parasitic capacitances and RC delays should be back annotated to DC or PT, for final SDF generation.

The following commands may be used to back annotate the extracted data to the design and specify the clock information while generating the post-layout SDF file for simulation:

DC Commands

read_timing –format sdf <interconnect RC's in SDF format>
include –quiet <parasitic capacitances in set_load format>

create_clock –period 30 –waveform {0 15} CLK
set_clock_skew –propagated CLK

PT Commands

read_sdf <interconnect RC's in SDF format>
source <parasitic capacitances in set_load format>
read_parasitics <DSPF, RSPF or SPEF file for clocks + other critical nets>

create_clock –period 30 –waveform [list 0 15] [list CLK]
set_propagated_clock [get_clocks CLK]

10.2.3 Issues Related to Timing Checks

Sometimes, during simulation, unknowns (X's) are generated that cause the simulation to fail. These unknowns are generated due to the violation of setup/hold timing checks. Most of the time, these violations are real, however there are instances where a designer may want to ignore some violations related to parts of the design, but still verify others. This is generally unachievable, due to the simulator's inability to turnoff the X-generation on a selective basis.

Nearly all simulators provide capabilities to ignore the timing violations, generally for the whole design. They do not have the ability to ignore the timing violation for an isolated instance of a cell in the design. Due to this reason, designers are often forced to either modify the simulation library or live with the failed result.

Modifying the simulation library is also not a viable approach, since turning off the X-generation can only be performed on a cell. This cell may be instanced multiple times in the design. Turning off the X-generation for this cell will prevent the simulator from generating X's, for all the instances of the cell in the design. This is definitely not desired as it may mask the real timing problems lying elsewhere in the design.

For example, a design may contain multiple clock domains and the data traverses from one clock domain to the other through synchronization logic. Although, this logic will work perfectly on a manufactured device, it may cause hold-time violations when simulated. This will cause the simulation to fail for the design.

Another example is related to the type of methodology used for synthesis. Some designers prefer to fix the hold-time violations only after layout. Failing to falsify or remove the hold-time values from the pre-layout SDF file may cause the simulator to generate an X (unknown) for the violating flop. This X may propagate to the rest of the logic causing the whole simulation to fail.

To prevent these problems, one may need to falsify selectively, the value of the setup and hold-time constructs in the SDF file, for simulation to succeed. The SDF file is instance based (rather than cell based), therefore selective targeting of the timing checks is easily attained. Instead of manually removing the setup and hold-time constructs from the SDF file, a better way is to zero out the setup and hold-times in the SDF file, only for the violating flops, i.e., replace the existing setup and hold-time numbers with zero's. Back-annotating the zero value for the setup and hold-time to the simulator prevents it from generating unknowns (if both setup and hold-time is zero, there cannot be any violation), thus making the simulation run smoothly. The following **dc_shell** command may be used to perform this:

```
dc_shell> set_annotated_check 0 –setup –hold    \
                              –from REG1/CLK  \
                              –to REG1/D
```

☺ Note: A similar command also exists for PT.

10.2.4 False Delay Calculation Problem

This topic is covered in Chapter 4, but is included here for the sake of completeness.

The delay calculation of a cell is based upon the input transition time and the output load capacitance of a cell. The input transition time of a cell is evaluated, based upon the transition delay of the driving cell (previous cell). If the driving cell contains more than one timing arc, then the worst transition time is used, as input to the driven cell. This causes a major problem when generating the SDF file for simulation purposes.

Consider the logic shown in Figure 10-2. The signals, *reset* and *signal_a* are inputs to the instance U1. Let us presume that the *reset* signal is not critical, while the *signal_a* is the one that we are really interested in. The *reset* signal is a slow signal therefore the transition time of this signal is more compared to *signal_a*, which has a faster transition time. This causes, two transition delays to be computed for cell U1 (2 ns from A to Z, and 0.3 ns from B to Z). When generating SDF, the two values will be written out separately as part of the cell delay, for the cell U1. However, the question now arises, which of the two values does DC use to compute the input transition time for cell U2? DC uses the worst (maximum) transition value of the preceding gate (U1) as the input transition time for the driven gate (U2). Since the transition time of *reset* signal is more compared to *signal_a*, the 2ns value will be used as input transition time for U2. This causes a large delay value to be computed for cell U2 (shaded cell).

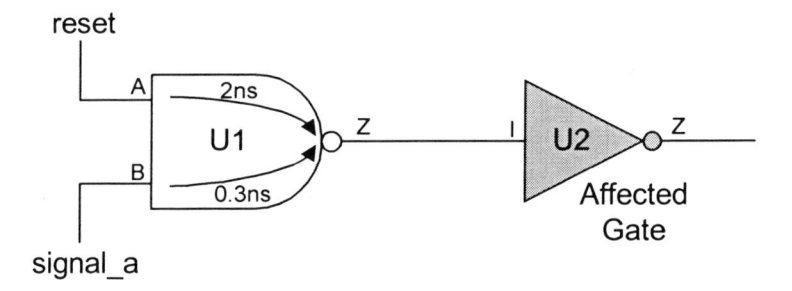

Figure 10-2. False Delay Calculation

To avoid this problem, one needs to instruct DC, not to perform the delay calculation for the timing arc, from pin A to pin Z of cell U1. This step should be performed before writing out the SDF. The following **dc_shell** command may be used to perform this:

```
dc_shell> set_disable_timing U1 –from A –to Z
```

SDF GENERATION

Unfortunately, this problem also exists during static timing analysis. Failure to disable the timing computation of the false path leads to large delay values computed for the driven cell.

10.2.5 Putting it Together

The following DC scripts combines all the information provided above and may be used to generate the pre and post-layout SDF, to be used for timing simulation of an example *tap controller* design.

DC script for pre-layout SDF generation

```
active_design = tap_controller

read active_design + ".db"

current_design active_design
link

set_wire_load LARGE –mode top
set_operating_conditions WORST

create_clock –period 33 –waveform {0 16.5} tck
set_clock_skew –delay 2.0  tck
set_clock_transition  0.2  tck

set_driving_cell –cell BUFF1 –pin Z all_inputs()
set_drive 0 {tck trst}

set_load 50 all_outputs()

set_input_delay   20.0 –clock tck –max all_inputs()
set_output_delay  10.0 –clock tck –max all_outputs()

/* Approximate the clock tree delay */
set_annotated_delay 2.0 –cell –from CLKPAD/A \
                              –to CLKPAD/Z
```

```
/* Assuming, only REG1 flop is violating hold-time */
set_annotated_check 0 –setup –hold   \
                    –from REG1/CLK   –to REG1/D

write_timing  –format sdf-v2.1   \
              –output active_design + ".sdf"
```

DC script for post-layout SDF generation

```
active_design = tap_controller

read active_design + ".db"

current_design active_design
link

set_operating_conditions BEST

include capacitance.dc   /* actual parasitic capacitances */
read_timing rc_delays.sdf  /* actual RC delays */

create_clock –period 33 –waveform {0 16.5} tck
set_clock_skew –propagated  tck

set_driving_cell –cell BUFF1 –pin Z all_inputs()
set_drive 0 {tck trst}

set_load 50 all_outputs()

set_input_delay    20.0 –clock  tck –max  all_inputs()
set_output_delay  10.0 –clock  tck –max all_outputs()
```

```
/* Assuming, only REG1 flop is violating hold-time */
set_annotated_check 0 –setup –hold      \
                    –from REG1/CLK   –to REG1/D

write_timing –format sdf-v2.1   \
             –output active_design + ".sdf"
```

10.3 Chapter Summary

The SDF file is used exhaustively throughout the ASIC world to perform dynamic timing simulations. The chapter briefly summarizes the contents of the SDF file that is related to ensuing discussions.

The chapter also discusses procedures for generating the SDF file from DC and PT, both for pre-layout and post-layout simulations. Along with command description, various helpful techniques are described to "massage" the SDF, in order for the simulation to succeed. These include fixing the clock latency and clock transition at the pre-layout level, and avoiding unknown propagation from selective logic of the design for successful simulation.

The final section gathered all the information and put it together in the form of DC scripts for pre and post-layout SDF generation.

11

PRIMETIME BASICS

PrimeTime (PT) is a sign-off quality static timing analysis tool from Synopsys. Static timing analysis or STA is without a doubt the most important step in the design flow. It determines whether the design works at the required speed. PT analyzes the timing delays in the design and flags violation that must be corrected.

PT, similar to DC, provides a GUI interface along with the command-line interface. The GUI interface contains various windows that help analyze the design graphically. Although the GUI interface is a good starting point, most users quickly migrate to using the command-line interface. Therefore, the intent of this chapter is to focus solely on the command-line interface of PT.

This chapter introduces to the reader, the basics of PT including a brief section devoted to Tcl language that is used by PT. Also described in this chapter are selected PT commands that are used to perform successful STA, and also facilitate the designer in debugging the design for possible timing violations.

11.1 Introduction

PT is a stand-alone tool that is not integrated under the DC suite of tools. It is a separate tool, which works alongside DC. Both PT and DC have consistent commands, generate similar reports, and support common file formats. In addition PT can also generate timing assertions that DC can use for synthesis and optimization. PT's command-line interface is based on the industry-standard language called Tcl. In contrast to DC's internal STA engine, PT is faster, takes up less memory, and has additional features.

11.1.1 Invoking PT

PT may be invoked in the command-line mode using the command pt_shell or in the GUI mode through the command primetime.

Command-line mode:
> pt_shell

GUI-mode:
> primetime

11.1.2 PrimeTime Environment

Upon invocation PT looks for a file called ".synopsys_pt.setup" and includes it by default. It first searches for this file in the current directory and failing that, looks for it in the users home directory before using the default setup file present in the PT installation site. This file contains the necessary setup variables defining the design environment that are used by PT, exemplified below:

set search_path   [list . /usr/golden/library/std_cells]

set link_path   [list {*} ex25_worst.db, ex25_best.db]

The variable search_path defines a list containing directories to look into while searching for libraries and designs. It saves the tedium of typing in complete file paths when referring to libraries and designs.

The variable link_path defines a list of libraries containing cells to be used for linking the design. These libraries are searched in the directories specified in the search_path. In the above example there are three elements in the list defined by link_path variable. The '*' indicates designs loaded in the memory, while the other two are names pertaining to the best and the worst case, standard cell technology libraries.

Another commonly used method of setting up the environment, if we do not want to use the ".synopsys_pt.setup" file is to use the source command. The source command works just like DC's include command. It includes and runs the file as if it were a script within the current environment. This command is invoked within pt_shell. For example:

```
pt_shell> source ex25.env
```

11.1.3 Automatic Command Conversion

Most of the DC commands are similar to PT commands, the exception being that PT being Tcl based uses the Tcl language format. This promotes the need for DC commands to be converted to the Tcl format before PT can utilize them.

PT offers a conversion script that may be used to convert almost all dc_shell commands to pt_shell, Tcl based format. The script is called `transcript` and is provided by Synopsys as a separate stand-alone utility. This script is executed from a UNIX shell as follows:

```
> transcript <dc_shell script filename> <pt_shell script filename>
```

11.2 Tcl Basics

Tcl provides most of the basic programming constructs – variables, operators, expressions, control flow, loops and procedures etc. In addition, Tcl also supports most of the UNIX commands.

Tcl programs are a list of commands. Commands may be nested within other commands through a process called command substitution.

Variables are defined and values assigned to them using the *set* command. For example:

>set clock_name   clk

>set clock_period 20

Values can be numbers or strings – Tcl does not distinguish between variables holding numeric values versus variables holding strings. It automatically uses the numeric value in an arithmetic context. In the above examples a variable *clock_name* has been set to the string *clk* and a variable *clock_period* has been set to the numeric value 20. Variables are used by preceding the variable name with a $. If the $ is not added before the variable name, Tcl treats it as a string.

>create_clock $clock_name –period 20 –waveform [0 10]

Arithmetic operations are performed through the expr command. This useful technique provides means for global parameterization across the entire script.

>expr $clock_period / 5

The above command returns a value 4. Tcl provides all the standard arithmetic operators like *, /, +, –, etc.

11.2.1 Command Substitution

Commands return a value that can be used in other commands. This nesting of commands is accomplished with the square brackets [and]. When a command is enclosed in square brackets, it is evaluated first and its value substituted in place of the command. For example:

>   set clock_period 20
>
>   set inp_del [expr $clock_period / 5]

The above command first evaluates the expression in the square brackets and then replaces the command with the evaluated value – in this case *inp_del* inherits the value of 4. Commands may also be nested to any depth. For example:

>   set clock_period 20
>
>   set inp_del [expr [expr $clock_period / 5] + 1]

The above example has 2 levels of nesting. The inner command returns a value of 4 obtained by dividing the value of *clock_period* by 5. The outer expr command adds 1 to the result of the inner command. Thus *inp_del* gets set to a value of 5.

11.2.2 Lists

Lists represent a collection of objects – these objects can be either strings or lists. In the most basic form enclosing a set of items in braces creates a list.

>   set clk_list {clk1 clk2 clk3}

In the above example, the set command creates a list called *clk_list* with 3 elements, *clk1, clk2* and *clk3*.

Another method of creating lists is through the list command that is typically used in command substitution. For example, the following list command creates a list just like the previous set command:

    list clk1 clk2 clk3

The list command is suitable for use in command substitution because it returns a value that is a list. The set command example may also be written as:

    set clk_list [list clk1 clk2 clk3]

Tcl provides a group of commands to manipulate lists – concat, join, lappend, lindex, linsert, list, llength, lrange, lreplace, lsearch, lsort and split. For example, concat concatenates two lists together and returns a new list.

    set new_list [concat [list clk1 clk2] [list clk3 clk4]]

In the above example the variable *new_list* is the result of concatenation of the list *clk1 & clk2* and the list *clk3 & clk4,* each of which is formed by the list command.

Conceptually a simple list is a string containing elements that are separated by white space. In most cases the following two are equivalent:

    {clk1 clk2 clk3}

    "clk1 clk2 clk3"

In some cases the second representation is preferred because it allows variable substitution while the first representation does not. For example:

    set stdlibpath [list "/usr/lib/stdlib25" "usr/lib/padlib25"]

    set link_path "/project/bigdeal/lib $stdlibpath"

For more details about the syntax of other list commands refer to any standard book on Tcl language.

11.2.3 Flow Control and Loops

Like other scripting and programming languages Tcl provides *if* and *switch* commands for flow control. It also provides *for* and *while* loops for looping. The *if* command may be used, along with *else* or *elsif* statements to completely specify the process flow. The arguments to *if*, *elsif* and *else* statements are usually lists, enclosed in braces to prevent any substitution. For example:

```
if {$port == "clk"} {
    create_clock –period 10 –waveform [list 0 5] $port
} elsif {$port == "clkdiv2"} {
    create_generated_clock –divide_by 2 –source clk $port
} else {
    echo "$port is not a clock port"
}
```

11.3 PrimeTime Commands

PT uses similar (but not same) commands as DC, to perform timing analysis and related functions. This is because of the difference in formats between DC and PT. A PT provided conversion script called *transcript* (explained earlier in the chapter) is used to convert the dc_shell commands to the PT equivalent. Since all relevant dc_shell commands are explained in detail in Chapter 6, comprehensive explanation is not provided in this section for all related commands.

11.3.1 Design Entry

Unlike DC, which can read RTL source files through HDL Compiler, PT being the static analysis engine can only read mapped designs. This performs the basis of design entry to PT. Among others, input to PT can be a file in db, Verilog, VHDL or EDIF format. The following pt_shell commands appropriate to each format are used to read the design in PT:

read_db –netlist_only <design name>.db   #db format

read_verilog <design name>.sv   #verilog format

read_vhdl <design name>.svhd   #vhdl format

read_edif <design name>.edf   #EDIF format

Since the netlist in db format can also contain constraints and/or environmental attributes (maybe saved by the designer), the –netlist_only option may be used for the read_db command to instruct PT to load only the structural netlist. This prevents PT from reading the constraints and/or other attributes associated with the design. Only the structural netlist is loaded.

11.3.2 Clock Specification

The concepts behind clock specification remains the same as the ones described for DC in Chapter 6. Subtle syntax differences exist due to difference in formats between the two. However, because clock specification may become complex, especially if there are internally generated clocks with clock division, this section will cover the complete PT clock specification techniques and syntax.

11.3.2.1 Creating Clocks

Primary clocks are defined as follows:

```
create_clock –period <value>
             –waveform {<rising edge> <falling edge>}
             <source list>
```

```
pt_shell> create_clock –period 20 –waveform {0 10}   \
                              [list CLK]
```

The above example creates a single clock named CLK having a period of 20ns, with rising and falling edges at 0ns and 10ns respectively.

11.3.2.2 Clock Latency and Clock Transition

The following commands are used to specify the clock latency and the clock transition. These commands are mainly used for pre-layout STA and are explained in detail in Chapter 12.

    set_clock_latency   <value> <clock list>

    set_clock_transition <value> <clock list>

pt_shell> set_clock_latency   2.5 [get_clocks CLK]

pt_shell> set_clock_transition 0.2 [get_clocks CLK]

The above commands define the clock latency for the CLK port as 2.5ns with a fixed clock transition value of 0.2ns.

11.3.2.3 Propagating the Clock

Propagating the clock is usually done after the layout tool inserts the clock tree in the design, and the netlist is brought back to PT for STA. The clock is propagated through the entire clock tree network in the netlist in order to determine the clock latency. In other words, the delay across each cell in the clock tree and the interconnect wiring delay between the cells is taken into account.

The following command instructs PT to propagate the clock through the clock network:

    set_propagated_clock <clock list>

pt_shell> set_propagated_clock [get_clocks CLK]

11.3.2.4 Specifying Clock Skew

Clock skew, or clock uncertainty as Synopsys prefers to call it, is the difference in the arrival times of the clock, at the clock pin of the flops. In synchronous designs data gets launched by the flop at one clock edge and is received by another flop at another clock edge (usually the next clock edge). If the two clock edges (launch and receive) are derived from the same clock then ideally there should be an exact delay of one clock period between the two edges. Clock skew puts a crimp in this happy situation. Because of variation in routing delays (or gated clock situation) the receiving clock edge may arrive early or late. Early arrival could cause setup-time violations and late arrival may cause hold-time violations. Therefore, it is imperative to specify the clock skew during the pre-layout phase, in order to produce robust designs.

Clock skew is specified through the following command:

```
set_clock_uncertainty  <uncertainty value>
                       –from <from clock>
                       –to <to clock>
                       –setup
                       –hold
                          <object list>
```

In the following example, 0.6ns is applied to both the setup and hold-time of the clock signal, CLK.

`pt_shell> set_clock_uncertainty 0.6 [get_clocks CLK]`

The option –setup may be used to apply uncertainty value to setup-time checks and while –hold option applies the uncertainty value for hold-time checks. It must be noted that different values for setup and hold cannot be implemented within a single command. Two separate commands must be used for this purpose. For example:

`pt_shell> set_clock_uncertainty 0.5 –hold [get_clocks CLK]`

`pt_shell> set_clock_uncertainty 1.5 –setup [get_clocks CLK]`

Also inter-clock skew can be specified with the −from and −to options, which is useful for designs containing multiple clock domains. For example:

```
pt_shell> set_clock_uncertainty 0.5 −from [get_clocks CLK1]   \
                                 −to [get_clocks CLK2]
```

11.3.2.5 Specifying Generated Clocks

This is an important feature that is absent from DC. Very often a design may contain internally generated clocks. PT allows the user to define the relationship between the generated clock and the source clock, through the command create_generated_clock. This is convenient because pre-layout scripts can be used for post-layout with minimal changes.

During post-layout timing analysis, clock tree is inserted and the clock latency is calculated by propagating the clock signal through the clock tree buffers. Users may opt to define the divided clock independent to the source clock (by defining the clock on an output pin of the dividing logic sub-block). However, this approach forces designers to manually add the clock tree delay (from the dividing block to the rest of the design) to the clock latency of the source clock to the dividing logic block.

By setting up a divided clock through the above command, the two clocks are kept in sync both in pre-layout and post-layout phases.

```
create_generated_clock −name <divided clock name>
                       −source <primary clock name>
                       −divide_by <value>
                        <pin name>

pt_shell> create_generated_clock −name DIV2CLK              \
                                 −source CLK −divide_by 2 \
                                 blockA/DFF1X/Q
```

The above example creates a generated clock on pin Q of the cell DFF1X belonging to blockA. The name of the generated clock is DIV2CLK, having half the frequency of the source clock, CLK.

11.3.2.6 Clock Gating Checks

For low power applications, designers often resort to gating the clock in the design. This technique allows designers to enable the clock only when needed. The gating logic may produce clipped clock or glitches, if the setup and hold-time requirements are not met (for the gating logic). PT allows designers to specify the setup/hold requirements for the gating logic, as follows:

>   set_clock_gating_check –setup <value>
>   –hold <value>
>   <object list>

pt_shell> set_clock_gating_check –setup 0.5 –hold 0.01 CLK

The above example informs PT that the setup-time and hold-time requirement for all the gates in the clock network of CLK is 0.5ns and 0.01ns respectively.

Gating checks on an isolated cell can be accomplished by specifying the cell name in the object list. For example:

pt_shell> set_clock_gating_check –setup 0.05 –hold 0.01   \
                         [get_lib_cell stdcell_lib/BUFF4X]

By default, PT performs the gating check with zero value used for setup and hold times – unless the library contains specific values for setup and hold times for the cell used to gate the clock. If the gating cell contains the setup/hold timing checks, then the gating check values may be automatically derived from the SDF file.

The clock gating checks are only performed for combinational cells. Also, the gating checks cannot be performed between two clocks.

11.3.3 Timing Analysis Commands

This section describes a selected set of PT commands that are used to perform STA. Only the most commonly used options are listed for these commands.

- **set_disable_timing**: Applications of this command include disabling timing arc of a cell in order to break the combinational feedback loop, or to instruct PT to exclude a particular timing arc (thus the path segment) from analysis.

    set_disable_timing –from <pin name>
                       –to <pin name>
                       <cell name>

  pt_shell> set_disable_timing –from A1 –to ZN {INVD2}

- **report_disable_timing:** command is used to display the timing arcs that were disabled by the user; or by PT. The report identifies individual disabled paths, using the following flags:

  Flags: u : Timing path disabled by the user.
           l : Timing loop broken by PT.
           c : Timing path disabled during case analysis.

- **set_input_transition**: is an alternative to the set_driving_cell command. It sets a fixed transition time is not dependent on the net loading. This command is specified on input/inout ports of the design.

    set_input_transition <value> <port list>

  pt_shell> set_input_transition 0.2 [all_inputs]

  pt_shell> set_input_transition 0.4 [list in1 in2]

- **set_timing_derate**: is used to derate the delay numbers shown in the timing report. PT provides this powerful capability that is useful in adding extra timing margin to the entire design. The amount of deration is controlled by a fixed value, which is specified by the user. The original delay numbers are multiplied by this value, before the timing report is generated.

    set_timing_derate –min <value> –max <value>

  pt_shell> set_timing_derate –min 0.2 –max 1.2

- **set_case_analysis**: command performs case analysis and is one of the most useful feature provided by PT. This command is used to set a fixed logic value to a port (or pin) while performing STA.

    set_case_analysis [0 | 1] <port or pin list>

  pt_shell> set_case_analysis 0 scan_mode

  Application of this command includes disabling timing paths that are not valid during a particular mode of operation. For instance, in the above example, the *scan_mode* port switches the design between the functional mode (normal operation) and the test mode of operation. The zero value set on the *scan_mode* port is propagated to all the cells driven by this port. This results in disablement of certain timing arcs of all cells that are related to the *scan_mode* port. Since testability logic is usually non timing-critical, disabling the timing arcs of the non timing-critical paths causes the real timing-critical paths to be identified and analyzed. The usage of this command is further explained in Chapter 12.

- **remove_case_analysis**: command is used to remove the case analysis values set by the above command.

    remove_case_analysis <port or pin list>

  pt_shell> remove_case_analysis scan_mode

- **report_case_analysis**: command is used to display the case analysis values set by the user. PT displays a report that identifies the pin/port list along with the corresponding case analysis values.

```
pt_shell> report_case_analysis
```

- **report_timing**: Similar to DC, this command is used to generate the timing report of path segments in a design. This command is used extensively and provides ample flexibility that is helpful in focussing explicitly on an individual path, or on a collection of paths in a design.

```
report_timing –from <from list>   –to <to list>
              –through <through list>
              –delay_type <delay type>
              –nets –capacitance –transition_time
              –max_paths <value> –nworst <value>
```

The –from and –to options facilitate the user in defining a path for analysis. Since there may be multiple paths leading from a startpoint to a single endpoint, the –through option may be used to further isolate the required path segment for timing analysis.

```
pt_shell> report_timing  –from [all_inputs]                    \
                         –to [all_registers –data_pins]

pt_shell> report_timing  –from in1                             \
                         –to blockA/subB/carry_reg1/D          \
                         –through blockA/mux1/A1
```

The –delay_type option is used to specify the type of delay to be reported at an endpoint. Accepted values are max, min, min_max, max_rise, max_fall, min_rise, and min_fall. By default PT uses the max type, which reports that the maximum delay between two points. The min type option is used to display the minimum delay between two points. The max type is used for analyzing the design for setup-time while the min type is used to perform hold-time analysis. The other types are not frequently used and users are advised to refer to the PT User Guide for full explanation regarding their usage.

```
pt_shell> report_timing  –from [all_registers –clock_pins]   \
                         –to [all_registers –data_pins]      \
                         –delay_type min
```

The –nets, –capacitance and –transition_time options are one of the most useful and frequently used options of the report_timing command. These options help the designer to debug a particular path, in order to track the cause of a possible violation. The –nets option displays the fanout of each cell in the path report, while the –capacitance and the –transition_time options reports the lumped capacitance on the net and the transition time (slew rate) for each driver or load pin, respectively. Failure to include these options results in a timing report that does not include the information mentioned above.

```
pt_shell> report_timing  –from in1                           \
                         –to blockA/subB/carry_reg1/D        \
                         –nets –capacitance –transition_time
```

The –nworst option specifies the number of paths to be reported for each endpoint, while the –max_paths option defines the number of paths to be reported per path group for different endpoints. The default value of both these options is 1.

```
pt_shell> report_timing  –from [all_inputs]                  \
                         –to [all_registers –data_pins]      \
                         –nworst 1000 –max_paths 500
```

- **report_constraint**: Similar to DC, this command in PT checks for the DRC's as defined by the designer or the technology library. Additionally, this command is also useful for determining the "overall health" of the design with regards to the setup and hold-time violations. The syntax of this command along with the most commonly used options is:

PRIMETIME BASICS

```
report_constraint  -all_violators  -max_delay
                   -max_transition  -min_transition
                   -max_capacitance  -min_capacitance
                   -max_fanout  -min_fanout
                   -max_delay  -min_delay
                   -clock_gating_setup  -clock_gating_hold
```

The –all_violators option displays all constraint violators. Generally, this option is used to determine at a glance, the overall condition of the design. The report summarizes all the violators starting from the greatest, to the least violator for a particular constraint.

`pt_shell>` report_constraint –all_violators

Selective reports may be obtained by using the –max_transition, –min_transition, –max_capacitance, –min_capacitance, –max_fanout, –min_fanout, –max_delay, and –min_delay options. The –max_delay and –min_delay options report a summary of all setup and hold-time violations, while others report the DRC violations. The –clock_gating_setup and the –clock_gating_hold commands are used to display the setup/hold-time reports for the cell used for gating the clock. In addition, there are other options available for this command that may be useful to the designer. Full details of these options may be found in the PT User Guide.

`pt_shell>` report_constraint –max_transition

`pt_shell>` report_constraint –min_capacitance

`pt_shell>` report_constraint –max_fanout

`pt_shell>` report_constraint –max_delay –min_delay

`pt_shell>` report_constraint –clock_gating_setup   \
                              –clock_gating_hold

☺ Initially use the report_constraint command to ascertain the amount, and the number of violations. The report produced provides a general estimate of the overall health of the design. Depending upon the severity of violations, a possible re-synthesis of the design may need to be performed. To further isolate the cause of the violation, the report_timing command should be used to target the violating path, in order to display a full timing report.

- **report_bottleneck**: This command is used to identify the leaf cells in the design that contribute to multiple violations. For instance, several violating path segments of a design may share a common leaf cell. Altering the size of this leaf cell (sizing up or down) may improve the timing (thus remove violation) of all the violating path segments. The syntax of this command along with the most commonly used options is:

```
report_bottleneck –from <from list> –to <to list>
                 –through <through list> –max_cells <value>
                 –max_paths <value> –nworst_paths <value>
```

The –from and –to options facilitate the user in defining a path for bottleneck analysis. Since there may be multiple paths leading from a startpoint to a single endpoint, the –through option may be used to further isolate the required path segment for bottleneck analysis.

```
pt_shell> report_bottleneck  –from in1                        \
                             –to blockA/subB/carry_reg1/D     \
                             –through blockA/mux1/A1
```

As the name suggests, the –max_cells option specifies the number of leaf cells to be reported. The default value is 20.

The –nworst_paths option specifies the number of paths to be reported for each endpoint, while the –max_paths option defines the number of paths to be reported per path group for different endpoints. The default value of both these options is 100.

```
pt_shell> report_bottleneck  –from in1                         \
                             –to blockA/subB/carry_reg1/D      \
                             –through blockA/mux1/A1           \
                             –max_cells 50                     \
                             –nworst_paths 500 –max_paths 200
```

11.3.4 Other Miscellaneous Commands

- **write_sdf**: command generates the SDF file that contains delays and timing checks for each instance in the design. PT uses the wire-load models to estimate the delays of cells during the pre-layout phase. For post-layout, PT uses the actual annotated delays (from the physical layout) while generating the SDF file. The syntax of this command along with the most commonly used options is:

```
write_sdf  –version 1.0 | 2.1
           –no_net_delays
           –no_timing_checks
           <sdf output filename>
```

Unless explicitly specified, by default PT generates the SDF file in SDF version 2.1 format.

The –no_net_delays option specifies that the interconnect delays (INTERCONNECT field in the SDF file) are not to be written out separately in the SDF file. In this case, they are included as part of the IOPATH delay of each cell. This option is mainly used during the pre-layout phase because of the fact that the interconnect delays are based upon the wire-load models. However, the interconnect delays after layout are real and are based on the routed design. Therefore, in general this option should be avoided while generating the post-layout SDF file.

```
pt_shell> write_sdf –no_net_delays top_prelayout.sdf

pt_shell> write_sdf top_postlayout.sdf
```

Specification of the –no_timing_checks option forces PT to omit the timing-checks section (TIMINGCHECK field) from the SDF file. As described in Chapter 10, the timing-checks section contains the setup/hold/width timing checks. This option is useful for generating the SDF file that may be used to validate, only the functionality of the design through dynamic simulation, without bothering to check for setup/hold/width timing violations. Once the design passes functional validation, full SDF (no –no_timing_checks option) may be generated.

```
pt_shell> write_sdf –no_timing_checks top_prelayout.sdf
```

- **write_sdf_constraints**: This command is similar to the write_constraints command in DC and performs the same function. It is used to generate the path timing constraints in SDF format, which is used by the layout tool to perform timing driven layout. The syntax of this command along with the most commonly used options is:

    write_sdf_constraints –version <1.0 | 2.1>
    　　　　　　　　–from <from list> –to <to list>
    　　　　　　　　–through <through list>
    　　　　　　　　–cover_design
    　　　　　　　　–slack_lesser_than <value>
    　　　　　　　　–max_paths <value> –nworst <value>
    　　　　　　　　<constraint filename>

Unless explicitly specified, by default PT generates the constraint file in SDF version 2.1 format. The –from, –to and –through options facilitate the user in specifying a particular path to be written to the constraint file.

The –nworst option specifies the number of paths to be written to the constraint file for each endpoint, while the –max_paths option defines the number of paths to be considered for each constraint group. The default value of both these options is 1. The default settings of these options usually suffice for most designs.

```
pt_shell> write_sdf_constraints  –from in1                    \
                                 –to blockA/subB/carry_reg1/D \
                                 –through blockA/mux1/A1      \
                                 tdl.sdf
```

The –**cover_design** option is used to generate just enough unique path timing constraints to cover the worst path for each path segment in the design. When specified, all other options such as, –nworst, –to, –from and –**through** are ignored. Although this option is recommended by Synopsys, it should be used judiciously as it may produce long run-times, especially for large designs.

```
pt_shell> write_sdf_constraints –cover_design tdl.sdf
```

An alternative is to use the –**slack_lesser_than** option that specifies that any path that has a slack value greater than the one specified is to be ignored. This means that a negative slack value for a path segment is considered to be most critical and has the highest priority. Thus all critical paths may be universally selected by specifying a low value for this option, hence will be written out to the constraint file. All high slack values (less critical paths) will be ignored.

```
pt_shell> write_sdf_constraints –slack_lesser_than 1.5 tdl.sdf
```

- **swap_cell**: This command may be used to replace an existing cell in the design with another, having the same pinout.

    swap_cell <cell list to be replaced> <new design>

For example, if a path is failing due to hold-time violation and in order to fix the timing violation, you want to see the effect on the reported slack, by sizing down a particular leaf cell in the path, without changing the netlist. In this case the **swap_cell** command may be used at the command line to replace the existing cell with another, containing the same pinout.

```
pt_shell> swap_cell {U1} [get_lib_cell stdcell_lib/AND2X2]
```

In the above example, the instance U1 (say a 2-input AND gate with 8X drive strength) in a design is replaced by the AND2X2 gate (2X drive strength) from the "stdcell_lib" technology library.

11.4 Chapter Summary

Static timing analysis is one of the most critical steps for the entire ASIC chip synthesis flow. This chapter provides an introduction to PrimeTime that included PrimeTime invocation and its environment settings.

PrimeTime is a stand-alone static timing analysis tool, which is based on the universally adopted EDA tool language, Tcl. A brief section is included on the Tcl language in context of PrimeTime, to facilitate the designer in writing PrimeTime scripts and building upon them to produce complex scripts.

The last section covers all relevant PrimeTime commands that may be used to perform static timing analysis, design debugging and writing delay information in SDF format. In addition, this section also covers topics on design entry and clock specification, both for pre-layout and post-layout.

12

STATIC TIMING ANALYSIS

Using PrimeTime

The key to working silicon usually lies in successful completion of static timing analysis performed on a particular design. PT is a stand-alone tool by Synopsys that is used to perform static timing analysis. It not only checks the design for required constraints that are governed by the design specifications, but also performs comprehensive analysis of the design. This capability makes STA one of the most important steps in the entire design flow and is used by many designers as a sign-off criterion to the ASIC vendor.

This chapter illustrates the part of the design flow where PT is utilized. It covers both the pre-layout and the post-layout phases of the ASIC design flow process.

STA is closely integrated with the overall synthesis flow, therefore parts of this chapter may contain some repetition from elsewhere in this book.

12.1 Why Static Timing Analysis?

Traditional methods of analyzing gate-level designs using dynamic simulation are posing a bottleneck for large complex designs. Today, the trend is towards incorporating system-on-a-chip (SoC), which may result in millions of gates per ASIC. Verifying such a design through dynamic simulation poses a nightmare to designers and may prove to be impossible due to long run-times (usually days and sometimes weeks). Furthermore, dynamic simulation relies on the quality and coverage of the test-bench used for verification. Only parts of the logic that are sensitized are tested while the remaining parts of the design remain untested. To alleviate this problem, designers now resort to other means of verification such as STA to verify the timing; and formal verification technique to verify the functionality of the gate-level netlist against the source RTL. However, comprehensive sets of test-benches are still needed to verify the functionality of the source RTL. Thus, dynamic simulation is needed to solely verify the functionality of the design at the RTL level. This results in considerable reduction in run-time.

The STA approach is infinitely fast compared to dynamic simulation and verifies all parts of the gate-level design for timing. Due to the similar nature of the synthesis and the STA engine, the static timing analysis is well suited for verifying synthesized designs.

12.1.1   What to Analyze?

In general, four types of analysis is performed on the design, as follows:

a) From primary inputs to all flops in the design.
b) From flop to flop.
c) From flop to primary output of the design.
d) From primary inputs to primary outputs of the design.

All four types of analysis can be accomplished by using the following commands:

```
pt_shell> report_timing   –from [all_inputs]                        \
                          –to [all_registers –data_pins]

pt_shell> report_timing   –from [all_registers –clock_pins]   \
                          –to [all_registers –data_pins]
```

```
pt_shell> report_timing   –from [all_registers –clock_pins]   \
                          –to [all_outputs]

pt_shell> report_timing   –from [all_inputs]                  \
                          –to [all_outputs]
```

Although, using the above commands is a cleaner method of generating reports for piping it to individual files for analysis, however, PT takes longer time to perform each operation. PT takes less time to generate the same results, if the following commands are used:

```
pt_shell> report_timing   –to [all_registers –data_pins]

pt_shell> report_timing   –to [all_outputs]
```

12.2 Timing Exceptions

In most designs there may be paths that exhibit timing exceptions. For instance, some parts of the logic may have been designed to function as multicycle paths, while others may simply be false paths. Therefore, before analyzing the design, PT must be made aware of the special behavior exhibited by these paths. PT may report timing violation for multicycle paths if they are not specified as such. Also, path segments in the design that are not factual, must be identified and specified as false paths, in order to prevent PT from producing the timing reports for these paths.

12.2.1 Multicycle Paths

PT by default, treats all paths in the design as single-cycle and performs the STA accordingly, i.e., data is launched from the driving flop using the first edge of the clock, and is captured by the receiving flop using the second edge of the clock. This means that the data must be received by the receiving flop within one clock cycle (single clock period). In the multicycle mode, the data may take more than one clock cycle to reach its destination. The amount of time taken by the data to reach its destination is governed by the multiplier value used in the following command:

set_multicycle_path <multiplier value>
            –from <from list> –to <to list>

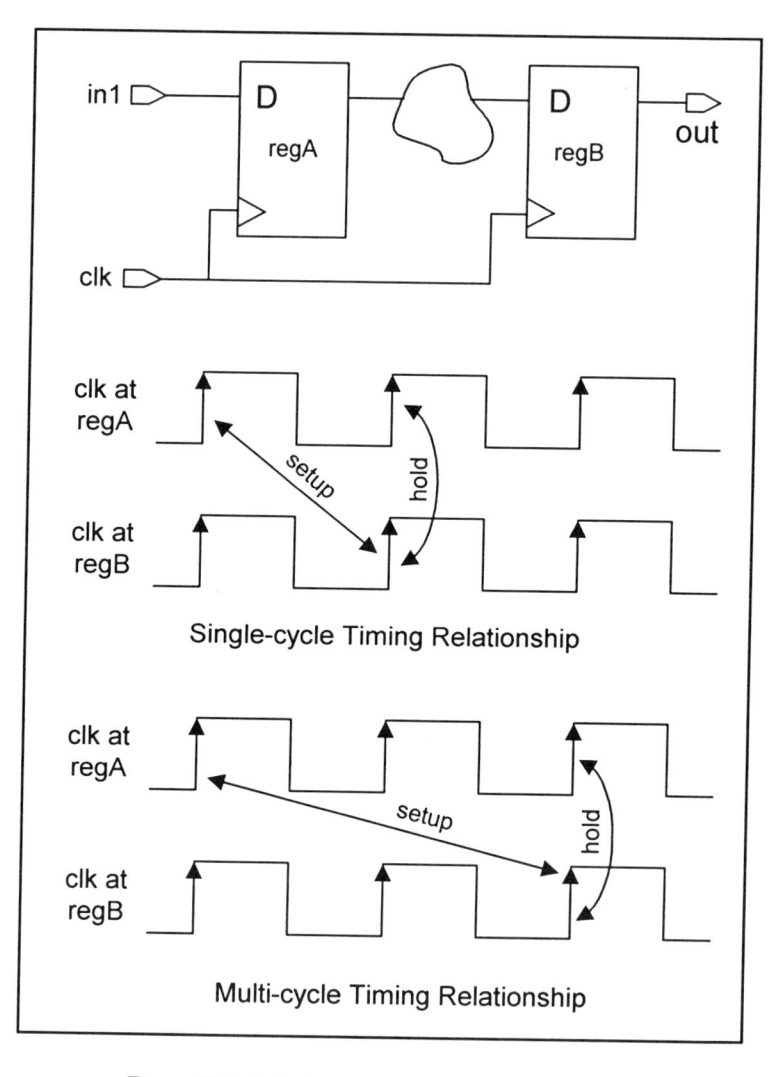

Figure 12-1. Defining Relationship for a Single Clock

Figure 12-1, illustrates the comparison between the single-cycle setup/hold-time relationship and the multicycle setup/hold-time relationship. In the multicycle definition, a multiplier value of 2 is used to inform PT that the data latching occurs at *regB* after an additional clock pulse. The following command was used:

```
pt_shell> set_multicycle_path 2 –from regA –to regB
```

In case of generated clocks, PT does not automatically determine the relationship between the primary clock and the derived clock, even when the create_generated_clock command is used. The single-cycle determination is independent of whether one clock is generated or not. It is based on the smallest interval between the open edge of the first clock to the closing edge of the second clock (in this case generated clock).

For separate clocks with different frequencies, the set_multicycle_path command may be used to define the relationship between these clocks. By default, PT uses the most restrictive setup-time and hold-time relationship between these clocks. These may be overridden by using the set_multicycle_path command that defines the exact relationship between these clocks.

Figure 12-2 illustrates an example, where a relationship exists between two separate clocks. During the single-cycle timing (default behavior), the setup and hold-time relationship occurs as shown. However, to specify a multicycle path between *regA* and *regB*, the following command is used:

```
pt_shell> set_multicycle_path 2 –setup    \
                    –from regA/CP –to regB/D
```

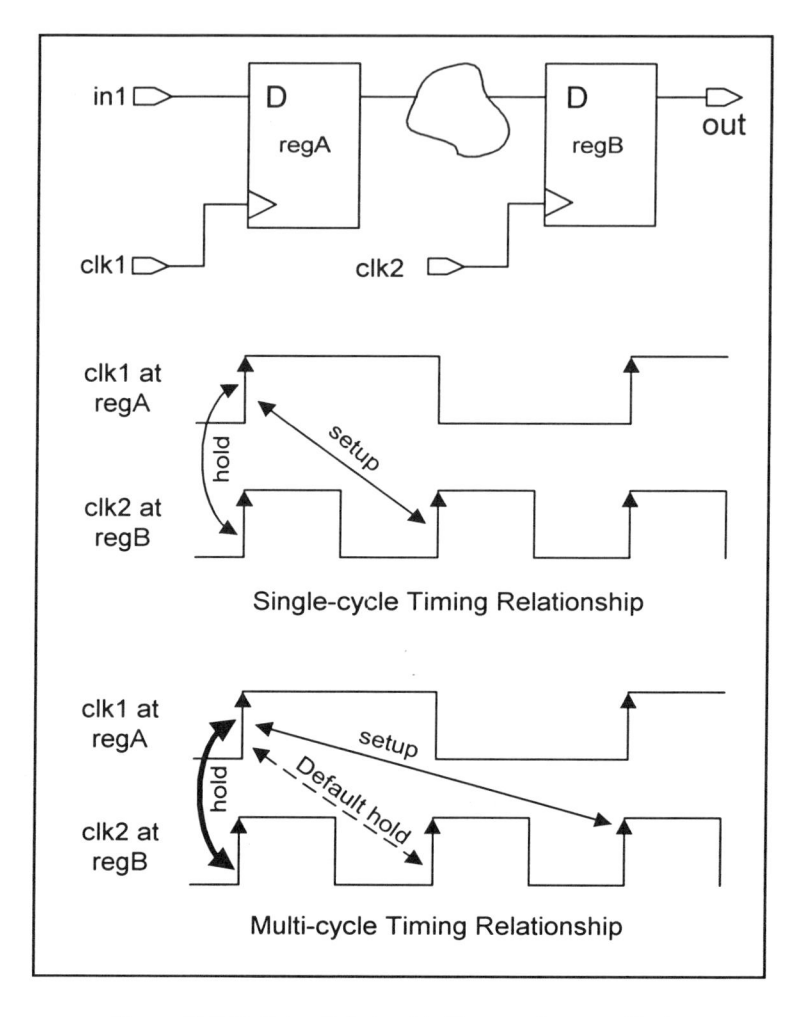

Figure 12-2. Defining Relationship Between Separate Clocks

The above example uses the multiplier value of 2 to define the setup-time relationship between the two clocks. The –setup option is used to define the setup-time relationship. However, this option also effects the hold-time relationship. PT uses a set of rules (explained in detail in PT User Guide) to determine the most restrictive relationship for the hold-time, between the two clocks. Therefore, PT may assume an incorrect hold-time relationship between the two clocks (shown as dotted line in Figure 12-2). To avoid this situation, the hold-time relationship between the two clocks should also be defined. Specification of the hold-time relationship through the set_multicycle_path command is very confusing, therefore not a recommended approach. Designers are advised to use the following command to specify the hold-time relationship between the two flops:

    `pt_shell>` set_min_delay 0 –from regA/CP –to regB/D

The zero value moves the hold-time relationship from the default value (dotted line in Figure 12-2) to the desired edge (bold line in Figure 12-2).

12.2.2 False Paths

Some designs may contain false timing paths. A false path is identified as a timing path that does not propagate a signal. False paths are created through the following **pt_shell** command:

    set_false_path –from <from list> –to <to list>
                    –through <through list>

It must be noted that the above command does not disable the timing arc of any cell, it merely removes the constraints of the identified path. Therefore, if the timing analysis is performed on the false path, an unconstrained timing report is generated.

By default, PT performs STA on all the paths. This results in the generation of timing reports for all the path segments (including the false paths in the design). If the false path segment is failing timing by a large amount then the report may mask the violations of the real timing paths. This of course depends upon the options used for the **report_timing** command.

Lets presume that there are multiple false paths in the design and they are all failing by a large amount during hold-time STA. However, the real timing paths are failing by a small margin. The false paths have not been identified because the user thinks that a large value of –nworst and –max_paths options will cover all the paths in the design (including the real timing paths), therefore identification of false paths is unnecessary. The user uses the following command to analyze the design:

```
pt_shell> report_timing –from [all_inputs]             \
                        –to [all_registers –data_pins]   \
                        –nworst 10000 –max_paths 1000   \
                        –delay_type min
```

The above method is certainly a viable approach and may not overly impact the run-time. However, a large value for the –nworst and –max_paths options (used in the above example) causes PT to generate/display multiple timing reports, covering all the paths in the design, most of which are false paths. Only a selected few timing reports relate to the real timing violations. By using this approach, it becomes tedious to distinguish between the real timing path and the false timing paths. In addition, due to the large amount of timing reports generated, it is easy to mistakenly overlook a real timing path that is violating the timing constraints. To avoid this situation, false path identification is recommended before performing STA.

In addition designers may use the –through option to further isolate the false path. It must be noted that the –through option significantly impacts the run-time, therefore should be used judiciously and the usage minimized. A better alternative is to disable the timing arc of the cell in the –through list, using the set_disable_timing command explained later in this chapter.

12.2.2.1 Helpful Hints for Setting False Paths

Timing exceptions impact the run-time. Setting multiple false paths in a design causes PT to slow down even further. Designers inadvertently specify the false paths with no regards to the proper usage, thereby impacting the run-time. The following suggestions are provided to help the designer in properly defining the false paths:

a) Avoid using wildcard characters when defining false path. Failing to do so may result in PT generating a large number of false paths. For example:

```
pt_shell> set_false_path –from ififo_reg*/CP   \
                         –to ofifo_reg*/D
```

In the above case, if the *ififo_reg* and *ofifo_reg* are each part of a 16-bit register bank, PT will generate a large number of unnecessary false paths. Disabling the timing arc of a common cell that is shared by the above paths is a better approach. The timing arc is disabled using the **set_disable_timing** command, explained in the next section.

b) Avoid using –through option for multiple false paths. Try finding a common cell that is shared by a group of identified false paths. Disable the timing arc of this cell through the **set_disable_timing** command.

c) Do not define false paths for registers belonging to separate asynchronous clock domains. For instance, if there are two asynchronous clocks (say, CLK1 and CLK2) then the following command should be avoided:

```
pt_shell> set_false_path –from [all_registers –clock CLK1] \
                        –to [all_registers –clock CLK2]
```

The above command forces PT to enumerate every register in the design, thereby causing a big impact on the run-time. A superior alternative is to set the false paths on the clocks itself, rather than the registers. Doing this prevents PT from enumerating all the registers in the design, therefore little or no impact on the run-time is observed. This is a preferred and efficient method of defining the asynchronous behavior of two clocks in PT. For example:

```
pt_shell> set_false_path –from [get_clocks CLK1]   \
                        –to [get_clocks CLK2]

pt_shell> set_false_path –from [get_clocks CLK2]   \
                        –to [get_clocks CLK1]
```

12.3 Disabling Timing Arcs

PT automatically disables timing paths that cause timing loops, in order to complete the STA on a design. However, designers sometimes find it necessary to disable other timing paths for various reasons, most prevalent being the need for PT to choose the correct timing path at all times. The timing arcs may be disabled by individually disabling the timing arc of a cell, or by performing case analysis on an entire design.

12.3.1 Disabling Timing Arcs Individually

During STA, sometimes it becomes necessary to disable the timing arc of a particular cell, in order to prevent PT from using that arc while calculating the path delay. The need to disable the timing arc arises from the fact that, in order to calculate the delay of a particular cell, PT uses the timing arc that produces the largest delay. This sometimes is undesired and produces false delay values. This is explained in detail in Chapter 4.

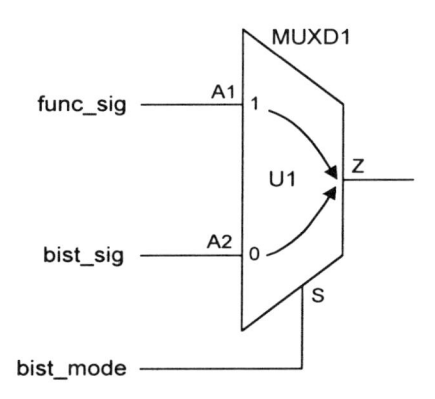

Figure 12-3. Disabling Timing Arcs

Another reason for disabling the timing arc of an individual cell is to prevent PT from choosing the wrong timing path. Figure 12-3 illustrates a case where the control input (*bist_mode*) is used to select between signals, *bist_sig* and *func_sig*, which are inputs to the multiplexer, MUXD1. The *bist_sig* signal is selected to propagate when the *bist_mode* signal is low, while the signal *func_sig* is allowed to pass when the *bist_mode* signal is high. During normal mode (functional mode), the signal *bist_sig* is blocked, while the signal *func_sig* is allowed to propagate. However, during test-mode (for e.g., for testing BIST logic), the *bist_sig* signal is selected to pass through, while blocking the *func_sig* signal. The application of this mux is described in detail in Chapter 8 (Figure 8-2), where it is used to bypass the input signals to the RAM, so that the logic previously shadowed by the RAM, thus unscannable, can be made scannable.

Three timing arcs exist for this cell – from A1 to Z, A2 to Z, and from S to Z. Only the first two arcs are shown in the above figure for the sake of clarity. While performing STA to check the timing in functional mode, unless the user isolates the path using the –through option of the report_timing command, PT may choose the wrong path (going through A2 to Z), thereby generating a false path delay timing report. Therefore it is prudent that the timing arc be disabled from A2 to Z of the cell MUXD1 (instance name U1) during functional mode STA. This is performed using the following pt_shell command:

pt_shell> set_disable_timing –from A2 –to Z {U1}

12.3.2 Case Analysis

An alternate solution to the above scenario is to perform case analysis on the design. By setting a logic value to the *bist_mode* signal, all timing arcs related to the *bist_mode* signal are disabled/enabled. In the above case, using the following command disables the timing arc from A2 to Z:

pt_shell> set_case_analysis 1 bist_mode

The logic 1 value for the *bist_mode* signal forces PT to disable the timing arc from A2 to Z and enables the signal *func_sig* to propagate. By changing this value to 0, the arc from A1 to Z is disabled and the *bist_sig* signal is allowed to propagate.

Although, both the set_disable_timing and set_case_analysis commands perform the same function of disabling the timing arcs, the case analysis approach is superior, for designs containing many such situations. For instance, a single command is used to analyze the entire design in either the normal mode or the test mode. However, the set_disable_timing command is useful for disabling the timing arc of an individual cell, when performing STA.

12.4 Environment and Constraints

Apart from slight syntax differences, the environment and constraints settings for PT are same as that used for DC. The following commands exemplify these settings:

pt_shell> set_wire_load_model –name <wire-load model name>

pt_shell> set_wire_load_mode < top | enclosed | segmented>

pt_shell> set_operating_conditions <operating conditions name>

pt_shell> set_load 50 [all_outputs]

pt_shell> set_input_delay   10.0 –clock <clock name> [all_inputs]

pt_shell> set_output_delay 10.0 –clock <clock name> [all_outputs]

Although, PT provides a multitude of options for the above commands, most designers only use a limited set of options, as shown above. Users are advised to refer to PT User Guide for full details regarding additional options available for each of the above commands.

Since the behavior and function of these commands are same as the commands used for DC, no explanation is given here. The DC commands that are related to each of the above command are explained in detail in Chapter 6.

12.4.1 Operating Conditions – A Dilemma

In general, the design is analyzed for setup-time violations utilizing the worst-case operating conditions, while the best-case operating condition is used to analyze the design for hold-time violations.

The reason for using the worst-case operating conditions to perform setup-time analysis is that the delay values of each cell in the library depict the delays (usually large) of a device operating under the worst-case conditions (maximum temperature, low voltage and other worst-case process parameters). The large delay values cause the data-flow to slow down, which may result in a setup-time failure for a particular flop.

An opposite effect occurs for the data-flow when the design uses the best-case operating conditions for hold-time STA. In this case, the delay values (small) of each cell in the technology library depict the best-case operating conditions (minimum temperature, high voltage and other best-case process parameters). Therefore, the data-flow now encounters less delay for it to reach its destination, i.e., the data arrives faster than before, which may cause hold-time violations at the input of the register.

By analyzing the design at both corners of the operating conditions, a time-window is created that states – if the device operates within the range defined by both operating conditions, the device will operate successfully.

12.5   Pre-Layout

After successful synthesis, the netlist obtained must be statically analyzed to check for timing violations. The timing violations may consist of either setup and/or hold-time violations.

The design was synthesized with emphasis on maximizing the setup-time, therefore you may encounter very few setup-time violations, if any. However, the hold-time violations will generally occur at this stage. This is due to the data arriving too fast at the input of sequential cells with respect to the clock.

If the design is failing setup-time requirements, then you have no other option but to re-synthesize the design, targeting the violating path for further optimization. This may involve grouping the violating paths or over-constraining the entire sub-block, which had violations. However, if the design is failing hold-time requirements, you may either fix these violations at the pre-layout level, or may postpone this step until after layout. Many designers prefer the latter approach for minor hold-time violations (also used here), since the pre-layout synthesis and timing analysis uses the statistical wire-load models and fixing the hold-time violations at the pre-layout level may result in setup-time violations for the same path, after layout. However, if the wire-load models truly reflect the post-routed delays, then it is prudent to fix the hold-time violations at this stage. In any case, it must be noted that gross hold-time violations should be fixed at the pre-layout level, in order to minimize the number of hold-time fixes, which may result after the layout.

12.5.1 Pre-Layout Clock Specification

In the pre-layout phase, the clock tree information is absent from the netlist. Therefore, it is necessary to estimate the post-route clock-tree delays up-front, during the pre-layout phase in order to perform adequate STA. In addition, the estimated clock transition should also be defined in order to prevent PT from calculating false delays (usually large) for the driven gates. The cause of large delays is usually attributed to the high fanout normally associated with the clock networks. The large fanout leads to slow input transition times computed for the clock driving the endpoint gates, which in turn results in PT computing unusually large delay values for the endpoint gates. To prevent this situation, it is recommended that a fixed clock transition value be specified at the source.

The following commands may be used to define the clock, during the pre-layout phase of the design.

```
pt_shell> create_clock –period 20 –waveform [list 0 10] [list CLK]

pt_shell> set_clock_latency    2.5 [get_clocks CLK]

pt_shell> set_clock_transition 0.2 [get_clocks CLK]

pt_shell> set_clock_uncertainty 1.2 –setup [get_clocks CLK]

pt_shell> set_clock_uncertainty 0.5 –hold  [get_clocks CLK]
```

The above commands specify the port CLK as type clock having a period of 20ns, the clock latency as 2.5ns, and a fixed clock transition value of 0.2ns. The clock latency value of 2.5ns signifies that the clock delay from the input port CLK to all the endpoints is fixed at 2.5ns. In addition, the 0.2ns value of the clock transition forces PT to use the 0.2ns value, instead of calculating its own. The clock skew is approximated with 1.2ns specified for the setup-time, and 0.5ns for the hold-time. Using this approach during pre-layout yields a realistic approximation to the post-layout clock network results.

12.5.2 Timing Analysis

The following script gathers all the information provided above and may be used to perform the setup-time STA on a design.

PT script for pre-layout setup-time STA

```
# Define the design and read the netlist only
    set active_design  <design name>

    read_db –netlist_only $active_design.db

# or use the following command to read the Verilog netlist.
# read_verilog $active_design.v

    current_design $active_design

    set_wire_load_model <wire-load model name>
    set_wire_load_mode  < top | enclosed | segmented >

    set_operating_conditions <worst-case operating conditions>

# Assuming the 50pf load requirement for all outputs
    set_load  50.0  [all_outputs]

# Assuming the clock name is CLK with a period of 30ns.
# The latency and transition are frozen to approximate the
# post-routed values.
```

```
create_clock –period 30 –waveform [0 15] CLK
set_clock_latency    3.0 [get_clocks CLK]
set_clock_transition 0.2 [get_clocks CLK]
set_clock_uncertainty 1.5 –setup [get_clocks CLK]
```

The input and output delay constraint values are assumed
to be derived from the design specifications.
```
set_input_delay    15.0 –clock CLK [all_inputs]
set_output_delay   10.0 –clock CLK [all_outputs]
```

Assuming a Tcl variable TESTMODE has been defined.
This variable is used to switch between the normal-mode and
the test-mode for static timing analysis. Case analysis for
normal-mode is enabled when TESTMODE = 1, while
case analysis for test-mode is enabled when TESTMODE = 0.
The bist_mode signal is used from the example illustrated in
Figure 12-3.

```
set TESTMODE [getenv TESTMODE]

if {$TESTMODE == 1} {
    set_case_analysis 1 [get_port bist_mode]
} else {
    set_case_analysis 0 [get_port bist_mode]
}
```

The following command determines the overall health
of the design.
```
report_constraint –all_violators
```

Extensive analysis is performed using the following commands.
```
report_timing –to [all_registers –data_pins]
report_timing –to [all_outputs]
```

Also, specification of the startpoint and the endpoint for the –from and the –to options of the report_timing command may be used to target selective paths. In addition, further isolation of the selected path may be achieved by using the –through option.

By default, PT performs the maximum delay analysis, therefore specification of the max value for the –delay_type option of the report_timing command is not needed. However, in order to display all timing paths of the design, the –nworst and/or –max_paths options may be utilized.

As mentioned in the previous chapter, the report_constraint command is used to determine the overall health of the design. This command should be initially used to check for DRC violations (max_transition, max_capacitance, and max_fanout etc.). In addition, this command may also be used to generate a broad spectrum of setup/hold-time timing reports for the entire design. Note that the timing report produced by the report_constraint command does not include a full path timing report. It only produces a summary report for all violating paths per endpoint (assuming that the –all_violators option is used).

The report_timing command is used to analyze the design in more detail. This command produces a timing report that includes the full path from the startpoint to the endpoint. This command is useful for analyzing the failing path segments of the design. For instance, it is possible to narrow down the cause of the failure, by utilizing the –capacitance and –net options of this command.

12.6 Post-Layout

The post-layout steps involve analyzing the design for timing with actual delays back annotated. These delays are obtained by extracting the layout database. The analysis is performed on the post-routed netlist that contains the clock tree information. Various methods exist for porting the clock tree to DC and PT, and have been explained in detail in Chapter 9. Let us assume that the modified netlist exists in db format.

At this stage, a comprehensive STA should be performed on the design. This involves analyzing the design for both the setup and hold-time requirements. In general, the design will pass timing with ample setup-time, but may fail hold-time requirements. In order to fix the hold-time violations, several methods may be employed. These are explained in Chapter 9. After incorporating the hold-time fixes, the design must be analyzed again to verify the timing of the fixes.

12.6.1 What to Back Annotate?

One of the most frequent questions asked by designers is – What should I back annotate to PT, and in what format?

Chapter 9, discusses various types of layout database extraction and associated formats. Pros and cons of each format are discussed at length. It is recommended that the following types of information be generated from the layout tool for back annotation to PT in order to perform STA:

a) Net RC delays in SDF format.

b) Capacitive net loading values in **set_load** format.

c) Parasitic information for clock and other critical nets in DSPF, RSPF or SPEF file formats.

The following PT commands are used to back annotate the above information:

- **read_sdf:** As the name suggests, this command is used read the SDF file. For example:

    ```
    pt_shell> read_sdf rc_delays.sdf
    ```

- **source:** PT uses this command to read external files in Tcl format. Therefore, this command may be used to back annotate the net capacitances file in **set_load** file format. For example:

    ```
    pt_shell> source capacitance.pt
    ```

- **read_parasitics:** This command is utilized by PT to back-annotate the parasitics in DSPF, RSPF and SPEF formats. You do not need to specify the format of the file. PT automatically detects it. For example:

    pt_shell> read_parasitics clock_info.spf

12.6.2 Post-Layout Clock Specification

Similar to pre-layout, the post-layout timing analysis uses the same commands, except that this time the clock is propagated through the entire clock network. This is because the clock network now comprises of the clock tree buffers. Thus the clock latency and skew is dependent on these buffers. Therefore, fixing the clock latency and transition to a specified value is not required for post-route clock specification. The following commands exemplify the post-route clock specification.

pt_shell> create_clock –period 20 –waveform [list 0 10] [list CLK]

pt_shell> set_propagated_clock [get_clocks CLK]

As the name suggests, the set_propagated_clock command propagates the clock throughout the clock network. Since the clock tree information is now present in the design, the delay, skew, and the transition time of the clock is calculated by PT, from the gates comprising the clock network.

12.6.3 Timing Analysis

Predominantly, the timing of the design is dependent upon clock latency and skew i.e., the clock is the reference for all other signals in the design. It is therefore prudent to perform the clock skew analysis before attempting to analyze the whole design. A useful Tcl script is provided by Synopsys through their on-line support on the web, called SolvNET. You may download this script and run the analysis before proceeding. If the Tcl script is not available, then designers may write their own script, to generate a report for the clock delay starting from the source point of the clock and ending at all the endpoints. The clock skew and total delay may be determined by parsing the generated report.

Although setting the clock uncertainty for post-layout STA is not needed, some designers prefer to specify a small amount of clock uncertainty, in order to produce a robust design.

Let us assume that the clock latency and skew is within limits. The next step is to perform the static timing on the design, in order to check the setup and hold-time violations. The setup-time analysis is similar to that performed for pre-layout, the only difference being the clock specification (propagate the clock) as described before. In addition, during post-route STA, the extracted information from the layout database is back annotated to the design.

The following script illustrates the process of performing post-route setup-time STA on a design. The items in bold reflect the differences between the pre and post-layout timing analysis.

PT script for post-layout setup-time STA

```
# Define the design and read the netlist only
    set active_design  <design name>

    read_db –netlist_only  $active_design.db

# or use the following command to read the Verilog netlist.
# read_verilog $active_design.v

    current_design  $active_design

    set_wire_load_model <wire-load model name>
    set_wire_load_mode  < top | enclosed | segmented >

# Use worst-case operating conditions for setup-time analysis
    set_operating_conditions <worst-case operating conditions>
```

```
# Assuming the 50pf load requirement for all outputs
    set_load  50.0  [all_outputs]

# Back annotate the worst-case (extracted) layout information.
    source capacitance_wrst.pt   #actual parasitic capacitances
    read_sdf rc_delays_wrst.sdf  #actual RC delays
    read_parasitics clock_info_wrst.spf  #clock network data

# Assuming the clock name is CLK with a period of 30ns.
# The latency and transition are frozen to approximate the
# post-routed values. A small value of clock uncertainty is
# used for the setup-time.
    create_clock –period 30 –waveform [0 15] CLK
    set_propagated_clock [get_clocks CLK]
    set_clock_uncertainty 0.5 –setup [get_clocks CLK]

# The input and output delay constraint values are assumed
# to be derived from the design specifications.
    set_input_delay    15.0 –clock CLK [all_inputs]
    set_output_delay 10.0 –clock CLK [all_outputs]

# Assuming a Tcl variable TESTMODE has been defined.
# This variable is used to switch between the normal-mode and
# the test-mode for static timing analysis. Case analysis for
# normal-mode is enabled when TESTMODE = 1, while
# case analysis for test-mode is enabled when TESTMODE = 0.
# The bist_mode signal is used from the example illustrated in
# Figure 12-3.

    set TESTMODE [getenv TESTMODE]

    if {$TESTMODE == 1} {
        set_case_analysis 1 [get_port bist_mode]
    } else {
        set_case_analysis 0 [get_port bist_mode]
    }
```

```
# The following command determines the overall health
# of the design.
    report_constraint –all_violators

# Extensive analysis is performed using the following commands.
    report_timing  –to [all_registers –data_pins]
    report_timing  –to [all_outputs]
```

As mentioned earlier, the design is analyzed for hold-time violations using the best-case operating conditions. The following script summarizes all the information provided above and may be used to perform the post-route hold-time STA on a design. The items in bold reflect the differences between the setup-time and the hold-time analysis.

PT script for post-layout hold-time STA

```
# Define the design and read the netlist only
    set active_design  <design name>

    read_db –netlist_only  $active_design.db

# or use the following command to read the Verilog netlist.
# read_verilog $active_design.v

    current_design  $active_design

    set_wire_load_model <wire-load model name>
    set_wire_load_mode  < top | enclosed | segmented >

# Use best-case operating conditions for hold-time analysis
    set_operating_conditions <best-case operating conditions>

# Assuming the 50pf load requirement for all outputs
    set_load  50.0  [all_outputs]
```

Back annotate the best-case (extracted) layout information.
    source capacitance_best.pt #actual parasitic capacitances
    read_sdf rc_delays_best.sdf #actual RC delays
    read_parasitics clock_info_best.spf #clock network data

Assuming the clock name is CLK with a period of 30ns.
The latency and transition are frozen to approximate the
post-routed values.
    create_clock –period 30 –waveform [0 15] CLK
    set_propagated_clock [get_clocks CLK]
    set_clock_uncertainty **0.2** –hold [get_clocks CLK]

The input and output delay constraint values are assumed
to be derived from the design specifications.
    set_input_delay   15.0 –clock CLK [all_inputs]
    set_output_delay 10.0 –clock CLK [all_outputs]

Assuming a Tcl variable TESTMODE has been defined.
This variable is used to switch between the normal-mode and
the test-mode for static timing analysis. Case analysis for
normal-mode is enabled when TESTMODE = 1, while
case analysis for test-mode is enabled when TESTMODE = 0.
The bist_mode signal is used from the example illustrated in
Figure 12-3.

    set TESTMODE [getenv TESTMODE]

    if {$TESTMODE == 1} {
        set_case_analysis 1 [get_port bist_mode]
    } else {
        set_case_analysis 0 [get_port bist_mode]
    }

The following command determines the overall health
of the design.
    report_constraint –all_violators

```
# Extensive analysis is performed using the following commands.
    report_timing  –to [all_registers –data_pins]          \
                           –delay_type min

    report_timing  –to [all_outputs] –delay_type min
```

12.7 Analyzing Reports

The following sub-sections illustrate the timing report generated by the report_timing command, both for pre-layout and post-layout analysis. A clock period of 30ns is assumed for the clock named *tck* of an example *tap_controller* design.

12.7.1 Pre-Layout Setup-Time Analysis Report

Example 12.1 illustrates the STA report generated during the pre-layout phase. The ideal setting for the clock is assumed using the pre-layout clock specification commands.

The following command was used to instruct PT to display a timing report for the worst path (maximum delay), starting at the input port *tdi* and ending at the input pin of a flip-flop.

```
pt_shell> report_timing –from tdi –to [all_registers –data_pins]
```

The default settings were used i.e., the –delay_type option was not specified, therefore PT performs the setup-time analysis on the design by assuming the **max** setting for the –delay_type option. Furthermore, PT uses the default values of –nworst and –max_paths options. This ensures that the timing report for a single worst path (minimum slack value) is generated. All other paths starting from the *tdi* input port and ending at other flip-flops will have a higher slack value, thus will not be displayed.

Example 12.1

```
*********************************************
Report   : timing
           –path full
           –delay max
           –max_paths 1
Design   : tap_controller
Version  : 1998.08–PT2
Date     : Tue Nov 17 11:16:18 1998
*********************************************

Startpoint: tdi (input port clocked by tck)
Endpoint: ir_block/ir_reg0
          (rising edge-triggered flip-flop clocked by tck)
Path Group: tck
Path Type: max
```

Point	Incr	Path	
clock tck (rise edge)	0.00	0.00	
clock network delay (ideal)	0.00	0.00	
input external delay	15.00	15.00	r
tdi (in)	0.00	15.00	r
pads/tdi (pads)	0.00	15.00	r
pads/tdi_pad/Z (PAD1X)	1.32	16.32	r
pads/tdi_signal (pads)	0.00	16.32	r
ir_block/tdi (ir_block)	0.00	16.32	r
ir_block/U1/Z (AND2D4)	0.28	16.60	r
ir_block/U2/ZN (INV0D2)	0.33	16.93	f
ir_block/U1234/Z (OR2D0)	1.82	18.75	f
ir_block/U156/ZN (NOR3D2)	1.05	19.80	r
ir_block/ir_reg0/D (DFF1X)	0.00	19.80	r
data arrival time		19.80	
clock tck (rise edge)	30.00	30.00	
clock network delay (ideal)	2.50	32.50	
ir_block/ir_reg0/CP (DFF1X)		32.50	r

library setup time	−0.76	31.74
data required time		31.74

---

data required time	31.74
data arrival time	− 19.80

---

slack (MET)	11.94

It is clear from the above report that the design meets the required setup-time with a slack value of 11.94ns. This means that there is a margin of at least 11.94ns before the setup-time of the endpoint flop is violated.

12.7.2 Pre-Layout Hold-Time Analysis Report

Example 12.2 illustrates the STA report generated during the pre-layout phase. The ideal setting for the clock is assumed using the pre-layout clock specification commands.

In order to perform hold-time STA, the following command was used to instruct PT to display a timing report for a minimum delay path, existing between two flip-flops.

```
pt_shell> report_timing –from [all_registers –clock_pins]   \
                        –to [all_registers –data_pins]      \
                        –delay_type min
```

In the above case, the –delay_type option was specified with min value, thus informing PT to display the best-case timing report. The default values of all other options were maintained.

Example 12.2

```
*******************************************
Report   : timing
         –path full
         –delay min
         –max_paths 1
```

STATIC TIMING ANALYSIS

```
Design   : tap_controller
Version  : 1998.08-PT2
Date     : Tue Nov 17 11:16:18 1998
*********************************************
```

Startpoint: state_block/st_reg9
       (rising edge-triggered flip-flop clocked by tck)
Endpoint: state_block/bp_reg2
       (rising edge-triggered flip-flop clocked by tck)
Path Group: tck
Path Type: min

Point	Incr	Path	
clock tck (rise edge)	0.00	0.00	
clock network delay (ideal)	2.50	2.50	
state_block/st_reg9/CP (DFF1X)	0.00	2.50	r
state_block/st_reg9/Q (DFF1X)	0.05	2.55	r
state_block/U15/Z (BUFF4X)	0.15	2.70	r
state_block/bp_reg2/D (DFF1X)	0.10	2.80	r
data arrival time		2.80	
clock tck (rise edge)	0.00	0.00	
clock network delay (ideal)	2.50	2.50	
state_block/bp_reg2/CP (DFF1X)		2.50	r
library hold time	0.50	3.00	
data required time		3.00	
data required time		3.00	
data arrival time		− 2.80	
slack (VIOLATED)		− 0.20	

A negative slack value in the above report implies that the hold-time of the endpoint flop is violated by 0.20ns. This is due to the data arriving too fast with respect to the clock.

To fix the hold-time for the above path, the setup-time analysis should also be performed on the same path in order to find the overall slack margin. Doing this provides a time frame in which the data can be manipulated.

For the above example, if the setup-time slack value is large (say, 10ns) then the data can be delayed by 0.20ns or more (say 1ns), thus providing ample hold-time at the endpoint flop. However, if the setup-time slack value is less (say 0.50ns) then a very narrow margin of 0.30ns (0.50ns minus 0.20ns) exists. Delaying the data by an exact amount of 0.20ns will produce the desired results, leaving 0.30ns as the setup-time. However, the minute time window of 0.30ns makes it extremely difficult for designers to fix the timing violation – delay the data just enough, so that it does not violate the setup-time requirements. In this case, the logic may need to be re-synthesized and the violating path targeted for further optimization.

12.7.3 Post-Layout Setup-Time Analysis Report

The same command that is used for pre-layout setup-time STA also performs the post-layout setup-time analysis. However, the report generated is slightly different, in the sense that PT uses asterisks to denote the delays that are back annotated.

Example 12.3 illustrates the post-layout timing report generated by PT to perform the setup-time STA. The same path segment shown in Example 12.1 (the pre-layout setup-time STA) is targeted to demonstrate the differences between the pre-layout and the post-layout timing reports.

Example 12.3

```
*********************************************
Report   : timing
           –path full
           –delay max
           –max_paths 1
Design   : tap_controller
Version  : 1998.08–PT2
Date     : Wed Nov 18 12:14:18 1998
*********************************************
```

Startpoint: tdi (input port clocked by tck)
Endpoint: ir_block/ir_reg0
          (rising edge-triggered flip-flop clocked by tck)
Path Group: tck
Path Type: max

Point	Incr	Path	
clock tck (rise edge)	0.00	0.00	
clock network delay (propagated)	0.00	0.00	
input external delay	15.00	15.00	r
tdi (in)	0.00	15.00	r
pads/tdi (pads)	0.00	15.00	r
pads/tdi_pad/Z (PAD1X)	1.30	16.30	r
pads/tdi_signal (pads)	0.00	16.30	r
ir_block/tdi (ir_block)	0.00	16.30	r
ir_block/U1/Z (AND2D4)	0.22 *	16.52	r
ir_block/U2/ZN (INV0D2)	0.24 *	16.76	f
ir_block/U1234/Z (OR2D0)	0.56 *	17.32	f
ir_block/U156/ZN (NOR3D2)	0.83 *	18.15	r
ir_block/ir_reg0/D (DFF1X)	1.03 *	19.18	r
data arrival time		19.18	
clock tck (rise edge)	30.00	30.00	
clock network delay (propagated)	2.00	32.00	
ir_block/ir_reg0/CP (DFF1X)		32.00	r
library setup time	−0.76	31.24	
data required time		31.24	
data required time		31.24	
data arrival time		− 19.18	
slack (MET)		12.06	

By comparison, the post-layout timing results improve from a slack value of 11.94 (in Example 12.1) to 12.06. This variation is attributed to the difference between the wire-load models used during pre-layout STA and the actual extracted back-annotated data from the layout. In this case, the wire-load models are slightly pessimistic as compared to the post-routed results.

Another difference between the pre-layout and the post-layout results is the propagation of the clock. In the pre-layout timing report, an ideal clock was assumed. However, during the post-layout STA the clock is propagated, thereby accounting for real delays. This is shown in the above report as "clock network delay (propagated)".

In the pre-layout phase, an ideal clock network delay of 2.5ns was assumed. The post-route STA results indicate that the clock is actually faster than previously estimated, i.e., the clock network delay value is 2.0ns instead of 2.5ns. This provides an indication to the post-routed clock network delay values. Therefore, the next time (next iteration, maybe) the design is analyzed in the pre-route phase, the clock network delay value of 2.0ns should be used to provide a closer approximation to the post-routed results.

12.7.4 Post-Layout Hold-Time Analysis Report

The same command that is used for pre-layout hold-time STA also performs the post-layout hold-time analysis. However, the report generated is slightly different, in the sense that PT uses asterisks to denote the delays that are back annotated. In addition, the clock network delay is propagated instead of assuming ideal delays.

Example 12.4 illustrates the post-layout timing report generated by PT to perform the setup-time STA. The same path segment shown in Example 12.2 (the pre-layout hold-time STA) is targeted to demonstrate the differences between the pre-layout and the post-layout timing reports.

Example 12.4

```
*********************************************
Report   : timing
         –path full
         –delay min
         –max_paths 1
```

STATIC TIMING ANALYSIS

```
Design  : tap_controller
Version : 1998.08-PT2
Date    : Tue Nov 17 11:16:18 1998
*******************************************
```

Startpoint: state_block/st_reg9
        (rising edge-triggered flip-flop clocked by tck)
Endpoint: state_block/bp_reg2
        (rising edge-triggered flip-flop clocked by tck)
Path Group: tck
Path Type: min

Point	Incr	Path	
clock tck (rise edge)	0.00	0.00	
clock network delay (propagated)	1.92	1.92	
state_block/st_reg9/CP (DFF1X)	0.00	1.92	r
state_block/st_reg9/Q (DFF1X)	0.18	2.10	r
state_block/U15/Z (BUFF4X)	0.04 *	2.14	r
state_block/bp_reg2/D (DFF1X)	0.06 *	2.20	r
data arrival time		2.20	
clock tck (rise edge)	0.00	0.00	
clock network delay (propagated)	1.54	1.54	
state_block/bp_reg2/CP (DFF1X)		1.54	r
library hold time	0.50	2.04	
data required time		2.04	
data required time		2.04	
data arrival time		− 2.20	
slack (MET)		0.16	

In the above case, the hold-time for the endpoint flop is met with a margin of 0.16ns to spare. Notice the difference in clock latency between the startpoint flop (1.92ns) and the endpoint flop (1.54ns). The difference in latency gives rise to the clock skew. Generally, a small clock skew value is acceptable, however a large clock skew may result in race conditions within the design.

The race conditions cause the wrong data to be clocked by the endpoint flop. Therefore, it is advisable to minimize the clock skew in order to avoid such problems.

12.8 Advanced Analysis

This section provides an insight to the designer to perform advanced STA on the design. Depending upon the situation, designers may analyze the design in detail, utilizing the concepts and techniques described in the following sections.

12.8.1 Detailed Timing Report

Often, in a design a path segment may fail setup and/or hold-time and it becomes necessary to analyze the design closely, in order to find the cause of the problem.

Consider the timing report shown in Example 12.2. The hold-time is failing by 0.20ns. In order to find the cause of the problem, the following command was used:

```
pt_shell> report_timing –from state_block/st_reg9/CP   \
                       –to state_block/bp_reg2/D       \
                       –delay_type min                 \
                       –nets –capacitance –transition_time
```

In the above command, additional options namely, –nets, –capacitance and –transition_time are used. Although the above command uses all three options concurrently, these options may also be used independently.

The timing report shown in Example 12.5 is identical to the one shown in Example 12.2, except that it uses the above command to produce the timing report that includes additional information on the fanout, load capacitance and the transition time.

Example 12.5

```
*******************************************
Report  :  timing
           -path full
           -delay min
           -max_paths 1
Design  :  tap_controller
Version :  1998.08-PT2
Date    :  Tue Nov 17 11:16:18 1998
*******************************************
```

Startpoint: state_block/st_reg9
        (rising edge-triggered flip-flop clocked by tck)
Endpoint: state_block/bp_reg2
        (rising edge-triggered flip-flop clocked by tck)
Path Group: tck
Path Type: min

Point	Fanout	Cap	Trans	Incr	Path
clock tck (rise edge)			0.30	0.00	0.00
clock network delay (ideal)				2.50	2.50
state_block/st_reg9/CP (DFF1X)			0.30	0.00	2.50 r
state_block/st_reg9/Q (DFF1X)			0.12	0.05	2.55 r
state_block/n1234 (net)	2	0.04			
state_block/**U15/Z (BUFF4X)**			0.32	**0.15**	2.70 r
state_block/n2345 (net)	**8**	**2.08**			
state_block/bp_reg2/D (DFF1X)			0.41	0.10	2.80 r
data arrival time					2.80
clock tck (rise edge)			0.30	0.00	0.00
clock network delay (ideal)				2.50	2.50
state_block/bp_reg2/CP (DFF1X)					2.50 r
library hold time				0.50	3.00
data required time					3.00
data required time					3.00

data arrival time		− 2.80
slack (VIOLATED)		− 0.20

By analyzing the timing report shown in Example 12.5, it can be seen that the cell U15 (BUFF4X) has a fanout of 8, with a load capacitance of 2.08pf. The computed cell delay is 0.15ns. As stated before, the hold-time violation is fixed by delaying the data with respect to the clock. Therefore, if the drive strength of the cell U15 is reduced from 4X to 1X, it will result in an increased delay value for the cell U15, due to the increase in transition time. This increase in delay value will contribute towards slowing the entire data path, thus removing the hold-time violation. The resulting timing report is shown in Example 12.6.

Example 12.6

```
*********************************************
Report  : timing
          −path full
          −delay min
          −max_paths 1
Design  : tap_controller
Version : 1998.08−PT2
Date    : Tue Nov 17 11:16:18 1998
*********************************************
```

Startpoint: state_block/st_reg9
    (rising edge-triggered flip-flop clocked by tck)
Endpoint: state_block/bp_reg2
    (rising edge-triggered flip-flop clocked by tck)
Path Group: tck
Path Type: min

Point	Fanout	Cap	Trans	Incr	Path
clock tck (rise edge)			0.30	0.00	0.00
clock network delay (ideal)				2.50	2.50

```
state_block/st_reg9/CP (DFF1X)           0.30   0.00   2.50 r
state_block/st_reg9/Q (DFF1X)            0.12   0.05   2.55 r
state_block/n1234 (net)         2  0.04
state_block/U15/Z (BUFF1X)               1.24   0.40   2.95 r
state_block/n2345 (net)         8  2.08
state_block/bp_reg2/D (DFF1X)            1.25   0.10   3.05 r
data arrival time                                      3.05

clock tck (rise edge)                    0.30   0.00   0.00
clock network delay (ideal)                     2.50   2.50
state_block/bp_reg2/CP (DFF1X)                         2.50 r
library hold time                               0.50   3.00
data required time                                     3.00
-----------------------------------------------------------------
data required time                                     3.00
data arrival time                                    − 3.05
-----------------------------------------------------------------
slack (MET)                                            0.05
```

In the timing report shown above, by reducing the drive strength of the cell U15 from 4X to 1X, an increase in transition time, and therefore an increase in the incremental delay of the gate is achieved. This impacts the overall data path, which results in a positive slack margin of 0.05ns, thus removing the hold-time violation for the endpoint flop.

12.8.2 Cell Swapping

PT allows the ability to swap cells in the design, as long as the pinout of the existing cell is identical to the pinout of the replacement cell. This capability allows designers to perform what-if scenarios without leaving the pt_shell session.

In Example 12.6, the cell BUFF1X (having a lower drive strength and identical pinout to BUFF4X) replaced the cell BUFF4X. However, the process of replacement was not discussed. There are two methods to achieve this. The netlist could be modified manually before performing STA; or designers may use the cell swapping capability of PT to perform the what-if

STA scenarios on the violating path segments, before manually modifying the netlist.

Manually modifying the netlist before performing the what-if scenarios is certainly a viable approach. However, it is laborious. For the case shown in Example 12.6, first the pt_shell session is terminated, then the netlist modified manually (BUFF4X replaced with BUFF1X), and finally the pt_shell session is invoked again to re-analyze the previously violating path segment (from st_reg9 to bp_reg2). If the modifications to the netlist do not produce the desired results (the path segment is still violating timing) then the whole process needs to be repeated. This approach is certainly tedious and wasteful.

A preferred alternative is to use the following command to replace the existing cell with another:

```
pt_shell> swap_cell {U15} [get_lib_cell stdcell_lib/BUFF1X]
```

The above command can be used within the pt_shell session to replace the existing cell BUFF4X (instanced as U15), with BUFF1X (from the "stdcell_lib" technology library), and the path segment re-analyzed to view the effect of the swapping. This provides a faster approach to debugging the design and visualizing the effect of cell swapping without terminating the pt_shell session.

Note that the cell swapping only occurs inside the PT memory. The physical netlist remains unmodified. If the path segment and the rest of the design passes STA, this modification should be incorporated in the netlist by modifying the netlist manually.

12.8.3 Bottleneck Analysis

Sometimes a design may contain multiple path segments that share a common leaf cell. If these path segments are failing timing then changing the drive strength (sizing it up or down) of the common leaf cell may remove the timing violation for all the path segments. PT provides the capability of identifying a common leaf cell that is shared by multiple violating path

STATIC TIMING ANALYSIS 267

segments in a design. This is termed as bottleneck analysis and is performed by using the **report_bottleneck** command.

In Example 12.2, a hold-time violation exists for the path segment starting from state_block/st_reg9 and ending at state_block/bp_reg2. However, the hold-time violation also exists (shown in Example 12.7) for the path segment starting from the same startpoint (state_block/st_reg9) but ending at a different endpoint, state_block/enc_reg0.

Example 12.7

```
****************************************
Report   : timing
           -path full
           -delay min
           -max_paths 1
Design   : tap_controller
Version  : 1998.08-PT2
Date     : Tue Nov 17 11:24:10 1998
****************************************

Startpoint: state_block/st_reg9
            (rising edge-triggered flip-flop clocked by tck)
Endpoint: state_block/enc_reg0
            (rising edge-triggered flip-flop clocked by tck)
Path Group: tck
Path Type: min

Point                                 Incr      Path
-----------------------------------------------------------
clock tck (rise edge)                 0.00      0.00
clock network delay (ideal)           2.50      2.50
state_block/st_reg9/CP (DFF1X)        0.00      2.50  r
state_block/st_reg9/Q (DFF1X)         0.05      2.55  r
state_block/U15/Z (BUFF4X)            0.15      2.70  r
state_block/enc_reg0/D (DFF1X)        0.07      2.77  r
data arrival time                               2.77
```

clock tck (rise edge)	0.00	0.00
clock network delay (ideal)	2.50	2.50
state_block/enc_reg0/CP (DFF1X)		2.50 r
library hold time	0.50	3.00
data required time		3.00
---	---	---
data required time		3.00
data arrival time		− 2.77
---	---	---
slack (VIOLATED)		− 0.23

In the timing report shown above, the hold-time violation is 0.23ns. Visual inspection of the two timing reports (Example 12.2 and 12.7) reveal that a single cell BUFF4X (instanced as U15) is common to both path segments (st_reg9 → bp_reg2, and st_reg9 → enc_reg0). Thus, reducing the drive strength of this cell may eliminate the hold-time violation for both the path segments.

However, this process involves careful visual inspection of all the path segments in the design in an effort to identify the common leaf cell between the startpoint and the endpoint of all the violating path segments. This method can be extremely tedious for a large number of path segments.

The recommended method of identifying a common leaf cell between the startpoint and the endpoint of all the violating path segments is to perform the bottleneck analysis. For the above case (in Example 12.7), the following command was used to identify the common leaf cell shared by the violating path segments.

   pt_shell> report_bottleneck

Example 12.8 illustrates a report that was generated by PT, identifying the cell U15 (BUFF4X) as the common leaf cell shared by the two path segments mentioned above.

Example 12.8

```
*********************************************
Report  : bottleneck
          -cost_type path_count
          -max_cells 20
          -nworst_paths 100
Design  : tap_controller
Version : 1998.08-PT2
Date    : Tue Nov 17 12:09:09 1998
*********************************************
```

Bottleneck Cost = Number of violating paths through cell

Cell	Reference	Bottleneck Cost
U15	BUFF4X	2.00

Once the cell has been identified, it can be swapped with another in order to fix the timing violation of multiple path segments. Once again, a complete STA should be performed on the entire design. Any required changes (due to cell swapping etc.) should be manually incorporated in the final netlist.

12.8.4 Clock Gating Checks

Usually, low power designs contain clocks that are enabled by the gating logic, only when needed. For such designs, the cell used for gating the clock should be analyzed for setup and hold-time violations, in order to avoid clipping of the clock.

The setup and hold-time requirements may be specified through the set_clock_gating_check command explained in Chapter 11. For example:

```
pt_shell> set_clock_gating_check -setup 0.5 -hold 0.02 tck
```

Example 12.9 illustrates the clock gating report that utilized the setup and hold-time requirements specified above for the gated clock, "tck". The following command was used to generate the report:

```
pt_shell> report_constraint –clock_gating_setup   \
                            –clock_gating_hold    \
                            –all_violators
```

Example 12.9

```
*******************************************
Report  : constraint
          –all_violators
          –path slack_only
          –clock_gating_setup
          –clock_gating_hold
Design  : tap_controller
Version : 1998.08–PT2
Date    : Tue Nov 17 12:30:07 1998
*******************************************

clock_gating_setup

Endpoint                                   Slack
---------------------------------------------------------------
state_block/U1789/A1                       –1.02 (VIOLATED)
state_block/U1346/A1                       –0.98 (VIOLATED)


clock_gating_hold

Endpoint                                   Slack
---------------------------------------------------------------
state_block/U1789/A1                       –0.10 (VIOLATED)
state_block/U1450/A1                       –0.02 (VIOLATED)
```

It is important to note that the –all_violators option should be used in addition to the –clock_gating_setup and the –clock_gating_hold options. Failure to include the –all_violators option will result in a report displaying only the cost function of the failures, instead of the identifying the failed gates.

The –verbose option may also be included to display a full path report for the purpose of debugging the cause of the violation, and how it may be corrected. Example 12.10 illustrates one such report that was generated by using the following command:

```
pt_shell> report_constraint –clock_gating_hold    \
                           –all_violators         \
                           –verbose
```

Example 12.10

```
*********************************************
    Report  : constraint
              –all_violators
              –path slack_only
              –clock_gating_hold
    Design  : tap_controller
    Version : 1998.08–PT2
    Date    : Tue Nov 17 12:32:10 1998
*********************************************

    Startpoint: state_block/tst_reg11
              (rising edge-triggered flip-flop clocked by tck)
    Endpoint: state_block/U1789
              (rising clock gating-check end-point clocked by tck)
    Path Group: **clock_gating_default**
    Path Type: min
```

Point	Incr	Path
clock tck (rise edge)	0.00	0.00
clock network delay (propagated)	2.25	2.25

```
state_block/tst_reg11/CP (DFF1X)    0.00              2.25  r
state_block/tst_reg11/Q (DFF1X)     0.05 *            2.30  r
state_block/U1789/A2 (AND4X)        0.12 *            2.42  r
data arrival time                                     2.42

clock tck (rise edge)               0.00              0.00
clock network delay (propagated)    2.50              2.50
state_block/U1789/A1 (AND4X)                          2.50  r
clock gating hold time              0.02              2.52
data required time                                    2.52
-----------------------------------------------------------
data required time                                    2.52
data arrival time                                   – 2.42
-----------------------------------------------------------
slack (VIOLATED)                                    – 0.10
```

In the above example, the AND0X gate is used to gate the clock, "tck". Pin A2 of this cell is connected to the enabling signal, whereas the clock drives pin A1 of this cell. As can be seen from the report, the hold-time is being violated by the gating logic. In order to fix the hold-time violation, the cell AND0X may be sized down to slow the data path.

12.9 Chapter Summary

Static timing is key to success, with working silicon as final product. Static timing not only verifies the design for timing, but also checks all the path segments in the design. This chapter covers all the steps necessary to analyze a design comprehensively through static timing analysis.

The chapter started by comparing static timing analysis to the dynamic simulation method, as the tool for timing verification. It was recommended that the former method be used as an alternative to dynamic simulation approach. This was followed by a detailed discussion on timing exceptions, which included multicycle and false paths. Helpful hints were provided to guide the user in choosing the best approach.

A separate section was devoted to disabling the timing arcs of cells and to perform case analysis. The case analysis was recommended over individual disabling of timing arcs, for designs containing many cells with timing arcs that are related to a common signal. An example case of DFT logic was provided as an application of case analysis.

In addition, the process of analyzing designs both for pre-layout and post-layout was covered in detail, which included clock specification and timing analysis.

Finally, a comprehensive section was devoted to timing reports followed by advanced analysis of the timing reports. At each step, example reports were provided and explained in detail.

Appendix

Example Makefile

```
# ==========================================
#  General Macros
# ==========================================

        MV=\mv -f
        RM=\rm -f
        DC_SHELL=dc_shell

        CLEANUP = $(RM) command.log pt_shell_command.log

# ==========================================
# Modules for Synthesis
# ==========================================

top:
        @make $(SYNDB)/top.db
$(SYNDB)/top.db:     $(SCRIPT)/top.scr     \
                     $(SRC)/top.v          \
                     $(SYNDB)/A.db         \
                     $(SYNDB)/B.db
$(DC_SHELL) -f  $(SCRIPT)/top.scr | tee top.log
```

```
$(MV)    top.log   $(LOG)
$(MV)    top.sv    $(NETLIST)
$(MV)    top.db    $(SYNDB)
$(CLEANUP)

A:
        @make $(SYNDB)/A.db
$(SYNDB)/A.db:      $(SCRIPT)/A.scr    \
                    $(SRC)/A.v         \
                    $(SYNDB)/A.db
$(DC_SHELL) -f $(SCRIPT)/A.scr | tee A.log
$(MV)    A.log     $(LOG)
$(MV)    A.sv      $(NETLIST)
$(MV)    A.db      $(SYNDB)
$(CLEANUP)

B:
        @make $(SYNDB)/B.db
$(SYNDB)/B.db:      $(SCRIPT)/B.scr    \
                    $(SRC)/B.v         \
                    $(SYNDB)/B.db
$(DC_SHELL) -f $(SCRIPT)/B.scr | tee B.log
$(MV)    B.log     $(LOG)
$(MV)    B.sv      $(NETLIST)
$(MV)    B.db      $(SYNDB)
$(CLEANUP)
```

Index

.synopsys_dc.setup, 45, 107
.synopsys_dc.setup file, 17
.synopsys_pt.setup, 45, 210
.synopsys_pt.setup file, 18

A

all_inputs, 232
all_outputs, 233
all_registers, 232
all_violators, 225, 271
analyze command, 55
area constraints, 126
area optimization, 145
ASIC design flow, 2
assign statements, 165
asterisks, 258, 260
ATPG, 151
Attributes, 51
Awk, 192

B

back annotated delays, 258, 260
back annotation to PT, 248
balance_buffer, 144
balanced clock tree, 173
balanced_tree, 66

Behavior Compiler, 5
Behavioral level, 5
best_case_tree, 67
BIST, 148
blocking assignments, 96
boolean, 138
bottleneck analysis, 226, 267
bottom-up compile, 16
boundary scan, 148
braces, 215
brute force method, 177
buffer_constants, 166
bus_naming_style, 18, 163
bypassing RAMs, 156

C

capacitance, 224
case analysis, 222, 240, 241
cell swapping, 265
cell_footprint, 186
change_names, 163
characterize, 133
characterize-compile, 130
characterize-write-script, 85
check_design, 164
check_test, 150
clock gating, 142, 220, 269
clock gating logic, 83
clock latency, 25, 174, 201, 217, 249

clock logic, 83
clock network, 25
clock propagation, 217, 260
clock skew, 143, 155, 173, 218, 249
clock skew analysis, 36, 249
clock skews, 113
clock spines, 113
clock transition, 25, 199, 217, 244
clock tree, 162
clock tree buffers, 249
clock tree synthesis, 30, 143, 173
clock tree transfer, 176
clock uncertainty, 18, 115, 218, 250
clocked scan, 149
clocking issues, 173
cluster, 171, 185
command substitution, 213
compile
   in_place, 187
   incremental, 191
   map_effort, incremental_mapping, in_place, only_design_rule, no_design_rule, scan, 136
   scan, 150
compile flow, 122
compile_disable_area_opt_during_inplace_opt, 187
compile_ignore_area_during_inplace_opt, 187
compile_ignore_footprint_during_inplace_opt, 187
compile_new_boolean_structure, 140
compile_ok_to_buffer_during_inplace_opt, 144, 187, 189
concat, 214
congestion, 178
connect_net, 177, 193
cost functions, 135

cover_design, 169, 229
create_cell, 177, 193
create_clock, 106, 114, 115, 117, 216, 244, 249
create_generated_clock, 219, 235
create_port, 177
create_test_patterns, 151, 152
create_wire_load, 185
critical paths, 126
critical range, 125
custom wire-load models, 35, 185, 187

D

dc_shell, 44
decoder, 152
default_max_fanout, 65
default_max_transition, 65
default_operating_conditions, 65
default_wire_load_mode
   top, enclosed, segmented, 69
default_wire_load_selection, 69
define_name_rules, 17, 163
delay calculation, 75, 203
delay calculator, 176
delay models, 73
derating, 222
Design Analyzer, 43
Design Compiler, 43
design constraints, 105
design entry, 215
design environment, 100
design objects, 48
design reuse, 80
design rule constraints, 104
design_analyzer, 44
detailed parasitics, 179
detailed route, 30, 181
detailed routing, 12, 178

DFT, 8, 147
directives, 56
disabling timing arcs, 221, 240
disconnect_net, 177, 193
dont_touch, 53, 84, 134, 162, 166
dont_use, 145
DRC, 35, 71, 129, 143, 151
DSPF, 179, 181, 248
dynamic simulation, 195, 232
Dynamic Simulation, 6

E

ECO, 13, 160
EDIF, 54, 160
elaborate command, 55
else statement, 215
elsif statement, 215
enclosed, 70, 102
enumerated types, 86
estimated delays, 181
expr command, 212
extraction, 179, 181, 248
extrapolate, 74

F

false paths, 110
false timing paths, 237
fanout, 262
fanout_length, 68
fanout_load, 71
fault coverage, 151
feedthroughs, 165
find, 52
finite state machines, 140
fishbone, 143
flat netlist, 176
flattening, 127, 137, 138, 141
flip-flops, 87

floorplan, 11
floorplanning, 30, 167
flow control, 215
for loop, 215
formal verification, 9, 29, 42, 176, 232
Formality, 43
forward annotating, 27, 168
full_case directive, 59

G

gated clocks, 112, 144, 153, 175
gated preset, 153
gated reset, 153
generated clocks, 116, 127, 129, 153, 219, 235
generating SDF, 26, 227
generating timing constraints, 228
generating timing reports, 223
generic statement, 55
get_attribute, 51
get_cells, 53
get_designs, 53
get_lib_cells, 53
get_nets, 53
get_ports, 53
global route, 30, 181
global routing, 12, 178
glue logic, 84
glue-logic, 80
group, 81
group_path, 113
grouping paths, 113, 125
GUI interface, 209

H

HDL, 5
hdlin_translate_off_skip_text, 61
hierarchical P & R, 168

hierarchical place and route, 36
hold, 114
HOLD timing check, 196
hold-time, 18, 183, 188, 237, 243
hold-time fixing, 189
hold-time static timing analysis, 252
hold-time violations, 39, 100

I

if command, 215
in_place, 136, 187
in_place_swap_mode, 65, 186
include, 184, 201
incremental_mapping, 136
in-place optimization, 39, 183, 186
input transition time, 221
insert_scan, 151
interconnect delay, 180
INTERCONNECT delay, 196
interpolation, 74
IOPATH delay, 196
IPO, 39, 183, 186

J

join, 214
JTAG, 8, 148

K

K-factors, 65

L

lappend, 214
latch inference, 88
latches, 87, 153
LBO, 39, 187
lbo_buffer_insertion_enabled, 189
lbo_buffer_removal_enabled, 189

lc_shell, 44
Library Compiler, 43
library group, 64
lindex, 214
link_library, 17, 46
link_path, 18, 46, 210
links to layout, 160
linsert, 214
list, 214
lists, 213
llength, 214
load capacitance, 262
location based optimization, 39, 187
logical hierarchy, 171
loops, 215
low power, 142
low power designs, 269
lrange, 214
lreplace, 214
lsearch, 214
lsort, 214
lssd, 149
LTL, 160
lumped parasitic capacitances, 179
lumped parasitics, 180
LVS, 35, 161, 176

M

makefile, 129
map_effort, 136
match_footprint, 65, 186
max_capacitance, 71
max_fanout, 71, 95
max_transition, 71
memory BIST, 8, 148, 156
memory elements, 87
minus_uncertainty, 114
multicycle, 233

multicycle paths, 110
multiple clock domains, 202
multiple clocks, 80, 84, 112, 127, 129
multiple false paths, 238
multiple path segments, 269
multiplexed flip-flop, 149
multiplexed_flip_flop, 18
mux inference, 93

N

negative slack, 257
net delays, 179
net loading, 248
nets, 224
no_design_rule, 136
non-blocking assignments, 96
non-linear delay model, 73

O

only_design_rule, 136
Open Verilog International, 179
operating conditions, 18, 66, 242
operating_conditions, 66
ordering scan-chains, 156
over constrain, 125
OVI, 179

P

parallel_case directive, 59
parasitic capacitances, 30
partitioning, 80
PATHCONSTRAINT, 169
PDEF, 171, 183, 187, 194
PDEF format, 34
percentile, 185
Perl, 192
physical design exchange format, 171

physical hierarchy, 171
pi model, 179
placement, 11
placement quality, 178
plus_uncertainty, 114
positive slack, 126
post-layout hold-time analysis, 260
post-layout optimization, 180, 183
post-layout SDF, 201
post-layout setup-time analysis, 258
pragma, 61
pre-layout hold-time analysis, 256
pre-layout SDF, 198
pre-layout setup-time analysis, 254
preview_scan, 150
primetime, 45, 210
PrimeTime, 43
priority encoder, 93
propagated, 115
pt_shell, 45, 210

R

race conditions, 96, 261
random logic, 140
RC delay, 174, 180, 181, 248
RC delays, 30, 143
read command, 55
read_clusters, 172, 184
read_db
  netlist_only, 216
read_edif, 216
read_parasitics, 185, 202, 249
read_sdf, 185, 202, 248
read_timing, 184, 201
read_verilog, 216
read_vhdl, 216
real delays, 181
reduced parasitics, 179

register inference, 90
Register Transfer Level, 5
remove_attribute, 53, 162, 166
remove_case_analysis, 222
remove_unconnected_ports, 164
reoptimize_design, 189
  in_place, 187
reoptimize_design_changed_list_file_name, 189
report_bottleneck, 226, 267, 268
report_case_analysis, 223
report_constraint, 225
  clock_gating_setup,
    clock_gating_hold, 270
report_disable_timing, 221
report_net, 144
report_test, 151
report_timing, 119, 223, 232, 233, 238, 246, 252, 254
  nets, capacitance, transition_time, 224, 262
report_transitive_fanout, 166
reporting DRCs, 224
routing, 11
RSPF, 179, 181, 248
RTL, 82
rules, 163

S

scaling factors, 65
scan, 136
scan insertion, 8, 21, 84, 149
scan style, 150
SDF, 169, 179, 181, 195, 248
SDF file, 116
SDF file generation, 198
search_path, 17, 18, 46, 210
segmented, 70, 102

sensitivity lists, 86
separate clocks, 235
set command, 212
set_annotated_check, 203
set_annotated_delay, 200
set_attribute, 51, 73
set_case_analysis, 222, 241, 246, 251, 253
set_clock_gating_check, 269
  setup, hold, 220
set_clock_latency, 199, 217, 244
set_clock_skew, 110, 114, 115
  delay, 199
  propagated, 201
set_clock_transition, 110, 114, 199, 217, 244
set_clock_uncertainty
  setup, hold, 218, 244
set_disable_timing, 76, 84, 204, 221, 241
set_dont_touch, 53, 107
set_dont_touch_network, 53, 106, 143, 144, 166
set_dont_use, 107
set_drive, 103
set_driving_cell, 103, 221
set_false_path, 111, 237, 238
set_fix_hold, 190, 191
set_fix_multiple_port_nets, 165
set_flatten, 138, 139
set_input_delay, 108, 190, 242
set_input_transition, 221
set_load, 52, 104, 179, 181, 242, 248
set_max_area, 126, 141
set_max_capacitance, 104
set_max_delay, 106, 112
set_max_fanout, 104
set_max_transition, 104
set_min_delay, 106, 113, 237

set_min_library, 100, 191
set_multicycle_path, 111, 234, 235
set_operating_conditions, 102, 242
set_output_delay, 109, 242
set_propagated_clock, 202, 217, 249
set_scan_configuration, 150
set_scan_element, 154
set_structure, 138, 141
set_timing_derate, 222
set_wire_load
   top, enclosed or segmented, 102
set_wire_load_mode
   top, enclosed, segmented, 242
set_wire_load_model, 242
setup, 114
setup files, 17
SETUP timing check, 196
setup-time, 18, 183, 236, 243
setup-time static timing analysis, 245
signal assignments, 97
single-cycle, 233, 235
SoC, 232
SolvNET, 249
source, 185, 202, 211, 248
spare gates, 13
SPEF, 179, 181, 248
spines, 143
split, 214
STA, 209
standard delay format, 195
state-machine, 80
static timing analysis, 10, 209
structuring, 138
   timing, boolean, 140
swap cells, 265
swap_cell, 229, 266
swapped, 269
swapping cells, 229

switch command, 215
symbol_library, 17, 46
synchronization logic, 84
synthesis environment, 7
synthesis_off directive, 62
synthesis_on directive, 62

T

target_library, 17, 46
Tcl, 212
test bench, 6
Test Compiler, 43
test ready compile, 8
test_default_scan_style, 18
test-bench, 195
test-ready compile, 149
time-budgeting, 128
timing driven layout, 27
timing exceptions, 233
timing range models, 67
timing reports, 254
timing_range, 67
TIMINGCHECK, 169
timing-driven layout, 168
timing-driven placement, 168
TNS, 125
top, 70, 102
top-down compile, 127
Total Negative Slack, 125
tran primitives, 165
transcript, 211
transition time, 262
transition_time, 224
translate_off directive, 57, 61
translate_on directive, 57, 61
transparent latches, 153
tree_type, 67
tri wires, 165

trim, 185
tri-state, 152
tri-state logic, 95
trunks, 143
two-pass synthesis, 190
types of analysis, 232

U

uncertainty, 114
ungroup, 81
    flatten, 142
uniquify, 134, 162
unknowns, 156, 202
unresolved references, 167
update_lib, 186

V

variable assignments, 97
Variables, 49
verbose, 271
Verilog, 5, 54, 160
verilogout_no_tri, 18, 165
verilogout_show_unconnected_pins, 18, 164

VHDL, 5, 54, 160
vhdlout_use_packages, 50
virtual clock, 106

W

while loop, 215
wire_load_selection, 69
wire-load, 20
wire-load models, 68, 171, 187
WNS, 125
Worst Negative Slack, 125
worst_case_tree, 67
write, 119
write_clusters, 172
write_constraints, 169
write_script, 130
write_sdf, 198, 227
write_sdf_constraints, 228
write_timing, 198, 206, 207

X

X-generation, 202

~~compile-preserve-sync-reset~~ — obsolete
set-fix-hold
prioritise-min-paths

0.2μ   total delay 70% wire

SPEF: Standard Parasitic Exchange Format

RSPF: Reduced Standard Parasitic Format (pi model)

DSPF: Detailed Standard Parasitic Format

printvar

Know A
Dr. Rajendra Prasad

MAPLE KIDS

KNOW ABOUT DR. RAJENDRA PRASAD

ALL RIGHTS RESERVED. No part of this book may be reproduced in a retrieval system or transmitted in any form or by any means electronic, mechanical, photocopying, recording and or without permission of the publisher.

Published by

MAPLE PRESS PRIVATE LIMITED
office: A-63, Sector 58, Noida 201301, U.P., India
phone: +91 120 455 3581, 455 3583
email: info@maplepress.co.in
website: www.maplepress.co.in

Printed in 2023, India

ISBN: 978-93-50334-07-2

Contents

Preface 4
1. Introduction - Childhood and education 6
2. His Meeting with Mahatma Gandhi 9
3. Freedom fighter 12
4. His Qualities 15
5. Becoming the Congress President 18
6. Becoming the First President of India 21
7. The Later Years of his Life 24
8. The Constitution of India 27
9. The Darbar 31
10. Election of the First President of India 34
11. The Ceremonial Procession 37
12. The Salient Features of the Constitution 40
13. Rajendra Prasad's Speech 43
14. His Respect for the Father of the Nation 48
15. His view on India at the Time of Independence 52
16. His Thoughts on Freedom 55
17. Words for Dr. Rajendra Prasad 58
18. Birth of National Flag 60
19. The Indian Legislature 65
20. The President House 69
21. Some Quotes on Dr. Rajendra Prasad 71

Preface

This book is the biography of Dr. Rajendra Prasad, the first President of India. He was born in Zeradei village, Siwan district, Bihar on December 3, 1884. In Zeradei's diverse population, people lived together in considerable harmony. Rajendra Prasad's earliest memories were playing 'Kabaddi' with his Hindu and Muslim friends.

Before entering politics, he taught English literature, history, economics, and law. In 1917, he began working with Mohandas K. Gandhi and in 1920, he joined the Indian National Congress and was several times appointed its President (1934, 1939, 1947-48). He was imprisoned (1942-45) for supporting Congress's opposition to the British war effort in World War II. He was one of those passionate individuals who gave up a lucrative profession to pursue a greater goal of attaining freedom for the Motherland.

After independence, Rajendra Prasad became the first President of India in 1950 and held that office until 1962. As President, he exercised his moderating influence.

On February 28, 1963, he passed away.

This book is an attempt to shed some light on the life of this great personality.

Chapter 1
Introduction - Childhood and Education

Dr. Rajendra Prasad was the son of Mahadev Sahay and was born on December 03, 1884, in Zeradei village, in the Saran district of North Bihar. Being the youngest in a large joint family, 'Rajen' was loved. He was attached to his mother and elder brother, Mahendra very much. Following the old customs of his village and family, Rajen was married when he was 12 years old, to Rajvanshi Devi.

Rajendra Prasad's great uncle, Chaudhuri Lal looked after the family with a zamindari income of Rs. 7,000 per year and substantial farmlands. He was the Deewan of the Hathwa Raj and was highly respected by all, honest, loyal and able persons. Rajendra Prasad's father, Mahadev Sahay, was a gentleman, a scholar of Persian and Sanskrit. His hobbies were wrestling, gardening, and providing free Ayurvedic and Unani treatment to patients who came to him. Rajendra Prasad's mother, Kamleshwari Devi, was a religious lady who never gave up her evening bath and *puja* even though she suffered from cough. Every day she would tell stories from the Ramayana to young Rajendra, as he slept close to her waiting for the morning light to peep into the windowless bedroom of their old-fashioned house. The Ramayana by Tulsidas became his faithful friend. He also loved to look at the Upanishads and other religious books.

The family shunned ostentations, lived and mixed freely with the co-villagers. There was a sense of community, fellow-feeling and kindness. All enjoyed festivals and the *pujas* together. The flow of village life was quiet and gentle. All this left a deep mark on young Rajendra's mind. The village stands for peace and rest.

At the age of five, young Rajendra according to the practice of the community was put under a Maulavi, who taught him Persian. Later, he was taught Hindi and Arithmetic. After completing this traditional education, he was put in the Chapra Zilla School. He moved to R. K. Ghosh's Academy in Patna to live with his only brother Mahendra Prasad, who was eight years older than him and joined Patna College. When Mahendra Prasad moved to Calcutta in 1897, Rajendra was admitted into the Hathwa High School. Soon, he rejoined the Chapra Zilla School, from where he passed the Entrance Examination of the Calcutta University at the age of eighteen, in 1902. Rajen was a brilliant student and stood first in the entrance examination of the University of Calcutta and was awarded a scholarship of Rs. 30 per month. He joined the famous Calcutta Presidency College in 1902. Gopal Krishna Gokhale formed 'The Servants of India Society' in 1905 and asked Rajen to join. But his duty towards his family and education was very high, so he refused Gokhale. But, this brought a sad change in him and for the first time in his life his academic performance went down.

After making his own choice, he kept the thoughts away and took his studies seriously. In 1915, Rajen passed the Masters in Law examination with honours, winning a gold medal. Later on he completed his Doctorate in Law as well.

Chapter 2
His Meeting with Mahatma Gandhi

Before he arrived in Calcutta he was introduced to the group of 'Swadeshi' by his elder brother Mahendra Prasad. The formation of the Bihari Students' Conference followed in 1908. It was the first organisation of this kind in India. It woke and practically produced the political leadership of the twenties in Bihar.

At the time, he set himself up as a lawyer in Calcutta in 1911. He was bound to Khan Bahadur Shamsul Huda, who joined the Indian National Congress and was elected to the All India Congress Committee. He impressed Sir Asutosh Mukherjee a year earlier that the latter offered him a Lectureship in the Presidency Law College. Gopal Krishna Gokhale, one of the greatest political leaders of India in those days, met him in Calcutta a year earlier and had asked him to join 'The Servants of India Society' in Poona. Still, the pressure of his family held him back and he started his practice in Patna. But he had no doubts about what he should do. He sought his elder brother Mahendra Prasad's permission to join Gokhale through a letter, which he had concluded with his innermost thoughts, which read, "Ambitions I have none, except to be of some service to the Motherland."

In April 1917, during the All India Congress Committee session, held in Calcutta, Gandhiji and Rajendra Prasad sat very close. Still, he did not know that Gandhiji was to be taken to his residence in Patna on his way to Champaran. This meeting with Gandhiji became a turning point in his career. He stayed with Gandhiji till his trial was over. Due to the Rowlatt Act and the Punjab disturbance. Thus, in 1920 before the civil disobedience and non-cooperation movement, he openly pledged to avoid unrighteous laws and follow the civil disobedience and non-cooperation. Thus, he constituted himself as an outlaw in the eyes of the British Government in India.

As an expert lawyer, Rajen realised that he would join the fight for independence at any time. While Gandhiji was on a fact-finding mission in the Champaran district of Bihar to feel the sorrow of the local farmers, he called Rajendra Prasad and asked him to come to Champaran with volunteers. Dr. Prasad rushed to Champaran. In the beginning, he did not like Gandhiji's looks or talks. But as time passed on, Dr. Prasad was deeply moved by the faithfulness, beliefs and courage that Gandhiji showed. He was a man who made the sufferings of the people of Champaran his own. Dr. Prasad decided that he would do everything to help them whether with his skills as a lawyer or as an active volunteer.

Chapter 3
Freedom Fighter

The ten years that followed were the years of great activity and heavy suffering. He took back his administrative post from the University to the regret of the British Vice-Chancellor. He withdrew his sons, Mrityunjaya and Dhananjaya, and his nephew, Janardan from the Benares Hindu University and other schools. He wrote articles for 'Searchlight' and the 'Desh' and collected funds for these papers. He toured a lot, explaining, lecturing and arguing. He was the first leading political figure in the Eastern Provinces, to join forces with Gandhiji, at a time when the latter was lacking large and effective followers. Another such leader from the West who joined Gandhiji was Vallabhbhai Patel. During the Nagpur Flag Satyagraha, Rajendra Babu and Vallabhbhai came closer. For Rajendra Babu, Sardar Vallabhbhai and his friends were very precious and it was one of the most pleasant memories of his life. He often went to Sabarmati and toured the country with Gandhiji. He suffered several terms of harsh imprisonment. He suffered from poverty, wanting a regular income of his own. All the while, he had asthma. He would not accept any financial help from Congress or any other source and mainly depended on his elder brother.

He was in jail, when on January 15, 1934, the destroying earthquake in Bihar occurred. He was released two days later. He was ill, yet he immediately started raising funds and organising relief. The Viceroy also raised a fund for the same purpose. His fund grew over 38 lakhs but the Viceroy's fund was one-third of the amount although he was known and honoured everywhere.

Nationalist India expressed its respect by electing Rajendra as the President of the Bombay session of the Indian National Congress. Mahendra Prasad, his elder brother was dead at that time. Congress remembered his social services and his loyalty towards the national cause through a formal resolution.

Gandhiji's influence changed many of Dr. Prasad's views, including caste and untouchability. Dr. Prasad reduced the number of servants to one and found ways to simplify his life. He no longer felt ashamed in sweeping the floor or washing his utensils and the works he always thought others would do.

When the Congress Ministries were formed in 1937, the Parliamentary Board consisted of Sardar Patel, Rajendra Babu and Maulana Azad. It provided effective guidance and control. In 1939, when Subhash Chandra Bose had to be relieved of the office of the Congress President, it was Rajendra Prasad who was persuaded to take over the President post and overcome the crises. Congress faced another difficulty when Acharya Kripalani resigned. Again, Rajendra Babu had to step into the branch. His management of the Constituent Assembly was excellent.

Chapter 4
His Qualities

Dr. Prasad was present to help people in pain. In 1914, floods affected Bihar and Bengal a lot. Dr. Prasad became a volunteer, distributing food and cloth to the flood victims.

In 1934, Bihar was shaken by an earthquake, which caused great damage and loss of property. The quake was terrible and was followed by floods and an outbreak of malaria, bringing misery. Dr. Prasad started with relief work, collecting food, clothes and medicine.

In 1935, an earthquake hit Quetta. Dr. Prasad was not allowed to help because of Government restrictions. He set up relief committees in Sind and Punjab for the homeless victims who came there.

Dr. Prasad called for non-cooperation in Bihar as part of Gandhiji's non-cooperation movement. Dr. Prasad gave up his law practice and started a National College near Patna in 1921. The college was later shifted to Sadaqat Ashram on the banks of the Ganga. The non-cooperation movement in Bihar spread like wildfire. Dr. Prasad toured the state, holding one public meeting after another, collecting funds and powering the nation for a complete boycott of schools, colleges and Government offices. He urged the people to spin and wear only Khadi.

Bihar and the whole nation rose, and the people responded to the leaders' call. The British Raj was coming to a stop.

The British India Government used the only option at its disposal. Mass arrests were made. Lala Lajpat Rai, Jawaharlal Nehru, Deshbandhu Chittranjan Das and Maulana Azad were arrested. Suddenly, the peaceful non-cooperation Movement turned to violence in Chauri Chaura, Uttar Pradesh. After the events at Chauri Chaura, Gandhiji suspended the civil disobedience movement. The entire nation was shocked. Gandhiji was criticised for what was called the 'Bardoli retreat'.

Chapter 5
Becoming the Congress President

Dr. Prasad stood by Gandhiji after seeing the wisdom behind Gandhiji's actions. Gandhiji did not want to set a precedence of violence for free India. In March 1930, Gandhiji launched the Salt Satyagraha. He planned to march from Sabarmati Ashram to Dandi seashore, to break the salt laws. A salt Satyagraha was launched in Bihar under Dr. Prasad. Nakhas Pond in Patna was chosen as the site of the Satyagraha. Batch after batch of volunteers was arrested while making salt. Many volunteers were injured. Dr. Prasad called for more volunteers. Public opinion forced the government to take back the police and allow the volunteers to make salt. Dr. Prasad then sold the manufactured salt to raise funds. He was sentenced to six months imprisonment.

His service to the freedom movement raised his status. Dr. Prasad presided over the Bombay session of the Indian National Congress in October 1934. After Subhash Chandra Bose resigned from the Presidentship of the Congress in April 1939, Dr. Prasad was elected the President. He did his best to unite the different ideologies of Subhash Chandra Bose and Gandhiji.

Rabindranath Tagore wrote to Dr. Prasad that his qualities would make both Gandhi and Bose forget their offended relationship and unite to bring peace in India.

As the freedom struggle went on, the shadow of communalism steadily grew in the background. To Dr. Prasad's dismay, communal riots began and burst all over the nation. He rushed from one place to another to control the riots. Independence was approaching fast and so was the chances of partition. Dr. Prasad, who had fond memories of playing with his Hindu and Muslim friends in Zeradei, witnessed the nation parted into two halves.

Chapter 6
Becoming the First President of India

In July 1946, when the Constituent Assembly was established to frame the Constitution of India, Dr. Rajendra Prasad was elected its President. Two and a half years after independence, on January 26, 1950, the Constitution of independent India was completed and Dr. Rajendra Prasad was elected the nation's first President. Dr. Prasad transformed the royal Rashtrapati Bhavan into a beautiful 'Indian' home. As a President, he used his moderating influence silently and unobtrusively and set a healthy precedent for others to follow. During his tenure as President, Dr. Prasad visited many countries to have good relationships, as the new state needs to establish and nourish new relationships. He stressed the need for peace in a nuclear age.

As President, he neither reined nor ruled but formed policies for the welfare and development of the state.

He never worried about what people said about him. He never looked back to history. There were occasions when he differed from the Prime Minister. But that was nothing new. They had differed for almost three decades yet worked together in the Congress. The differences never hampered their personal relations. Perhaps, both realised that they arose from their different backgrounds, beliefs, approaches and attitudes.

In 1960 when he announced his retirement many requested him to continue for a third term but his mind was made up. Jayaprakash Narayan was happy with his decision and suggested that he provide direct guidance to the Sarvodaya Movement after retirement. But the illness of 1961 shacked Rajendra Prasad's health completely. Many therefore, worried at his decision to go back to the

Sadaqat Ashram. Everyone worried about his guidance and the atmosphere of the ashram in which he would live. His elder sister Bhagwati Devi had passed away on January 25, 1960. She loved her younger brother dearly. She came to his house within two years of her marriage becoming a widow at nineteen years. On the next day, Rajendra Babu's strong willpower made him take the Republic Day salute as usual and calm. After returning from the parade, he takes his sister's body for cremation.

Chapter 7
The Later Years of his Life

After a few months of his retirement in September 1962, his wife Rajvanshi Devi passed away. Her contribution in making Rajendra as he was though indirect, was considerable. She was a pure, selfless and devoted wife. She asked for little and she had silently encouraged him and never stood in his way. Her husband's will and pleasure were hers. They used to sit silently for hours and they never talked to each other but they understood each other perfectly. Thus, his last days were days of sorrow.

The Chinese attack shook him completely. He had apprehended the danger. He thought of the worst conditions. But he was shaken away by the open attack. His will to live was becoming weak. In a letter to one devoted to him, he wrote it a month before his death: "I have a feeling that the end is near, end of the energy to do, end of my very existence." And so, when the end came suddenly on February 28, 1963, he was not unprepared. After a few hours' illnesses, he died with 'Ram Ram' on his lips.

Rajendra Babu shared Gandhiji's idea of making a new man in a new society. His mind was capable of thinking about broad issues in a short time. In 1962, after remaining President for 12 years, Dr. Prasad retired after which he was awarded the Bharat Ratna, the nation's highest civilian award. With many ups and downs of his vast life, Dr. Prasad recorded his life before and after independence in many books like *Satyagraha at Champaran* (1922), *India Divided* (1946), *Atmakatha* (1946), *Mahatma Gandhi and Bihar, Some Reminiscences* (1949) and *Bapu ke Kadmon Mein* (1954).

Here are some obvious facts about him:

A strict farmer who lived and dressed like them can impress an office where high standards and rich looks were more interesting! But he succeeded a lot.

Dr. Prasad spent the last few months of his life in retirement at the Sadaqat Ashram in Patna. He died on February 28, 1963.

India's first citizen was a man of dedication who always imagined possibilities to make life better and real.

Chapter 8
The Constitution of India

There were historic scenes of enjoying the beginning of Indian Independence on August 15, 1947, passing India's Constitution on November 26, 1949 and the introduction of the Republic of India, on January 26, 1950.

Fantastic scenes of enjoyment, celebrations and singing of national songs were seen on August 14-15, 1947. Gandhiji was then staying in Calcutta to promote his Peace Mission. Gandhiji's presence in riot-filled Calcutta created a miracle. Hindus and Muslims, who were unable to come out of their houses for months together because of riots, for the first time forgot all their differences and came out of their houses, greeted and hugged each other. The Mahatma saw the celebrations at night.

Loud and long cheers and beating on the desks greeted the passing of the Constitution in the Parliament House on November 26, 1949. The Constituent Assembly met in this hall. Several members shouted Vande Mataram and Bharat Mata Ki Jai in chorus and Dr. Rajendra Prasad signed the new Constitution of the Indian Republic.

In his speech, Dr. Rajendra Prasad, President of the Constituent Assembly, before passing the Constitution, paid honour to Mahatma Gandhi and said that those who in the future work with the Constitution must remember that unique methods taught by Gandhiji gained India's independence. They have to protect the Constitution and the state's independence by following non-violence and fearing God. He asked the people not to forget Mahatma Gandhi's teachings in their life.

The Constitution adopted by the Constituent Assembly of India represented the Government of the Republic of India. The Constitution signifies the sovereignty of the people of India. Our Constitution maintained democracy in India. The birth of our Republic marks the foundation for justice, liberty, equality, unity and integrity of the nation.

Prime Minister Jawaharlal Nehru congratulated Dr. Rajendra Prasad on passing the New Constitution.

After the Constitution was passed, the historic session of the Constituent Assembly ended with the singing of the National Anthem '*Jana-Gana-Mana -Adhinayaka Jai Hey, Bharat Bhagya Bidhata*', by Purnima Banerjee, a long-time freedom fighter and the sister of the late freedom fighter, Aruna Asaf Ali.

The Constituent Assembly elected Dr. Rajendra Prasad as President of the Indian Republic at a special session on January 24, 1950, agreeing with the Constitution.

On January 26, 1950, India became a Republic in the middle of the enjoyment and booming of guns. A notice that announced the new status of India was read out by the last Governor-General C. Rajagopalachari. It read that from January 26, 1950, India, that is Bharat, shall be the Sovereign Democratic Republic. It was a memorable and proud day for the whole nation.

Chapter 9
The Darbar

The high-domed circular Darbar Hall of Rashtrapati Bhavan (known as Government House) was brilliantly lit up. Over 500 guests had assembled inside the hall. President Sukarno of the Indonesian Republic, his wife and several members of the Diplomatic corps, members of the Constituent Assembly and prominent citizens graced the occasion. It was a historic occasion when free India's first President, Dr. Rajendra Prasad, took office oath. The Chief Justice of India, Sir Hiralal Kania, read the oath of office in Hindi. Dr. Rajendra Prasad repeated it sentence by sentence. The President was dressed in black *achkan*, white *churidar* and a white Gandhi cap.

The outgoing Governor-General, C. Rajagopalachari, the first Prime Minister, Pandit Jawaharlal Nehru, smiling with pride and joy, the Deputy Prime Minister, Sardar Patel (the 'Iron-man' of India), Cabinet Ministers, Judges of the Supreme Court and the Auditor-General of India, were present in the hall to witness this biggest national ceremony of the 20th century. Pandit Nehru and his other Cabinet Members took oath soon after. The Speaker of Lok Sabha, G. V. Mavalankar, the first Speaker, sat in the front row.

Outside the Darbar Hall, there were unforgettable scenes of Jubilation. Large crowds of men, women and children were seated in the open spaces of Rashtrapati Bhavan. Many of them had come from the nearby states to see the ceremony.

The capital wore the look of a national festival. People raised 'Gandhiji ki jai' and 'Vande Mataram' slogans. Thousands visited Rajghat, the Samadhi of Mahatma Gandhi, to pay their respectful homage to the 'Father of the Nation'.

In the Darbar Hall, for the first time, the national emblem of Ashoka Pillar with three lions was placed near the throne, where, in the past, the British Viceroys used to sit. Also, for the first time, a smiling statue of Lord Buddha was placed behind the throne.

The masses celebrated the birth of the Republic by organising 'Prabhat Pheries'. The day dawned with a clear sky and the sun was bright throughout the day. It was one of the coldest days in Delhi but men, women and children dressed in their best came out to participate in this great festival. They exchanged greetings and congratulated each other for the new era under the new Constitution.

Chapter 10
Election of the First President of India

Reproduced below is a report from a renowned newspaper on January 25, 1950, on the election of Dr. Rajendra Prasad as the President of India by the Constituent Assembly.

"Described by Sardar Patel as the 'red letter day', today saw the culmination of the Constituent Assembly's three years of labour. At today's sitting, it unanimously elected amidst deafening cheers, Dr. Rajendra Prasad as the first President of the Indian Republic, approved of the statement from the Chair regarding the national anthem and went through the ceremony of its members appending their signatures to three copies of the Constitution- two in English and one in Hindi.

After the Returning Officer, Mr. H. V. R. Iyengar, had announced the uncontested election of Dr. Rajendra Prasad. Pandit Nehru was the first to offer respectful congratulations and pledge 'my loyalty and fealty to this Republic of which you will be the head.'

Pandit Nehru paid a glowing tribute to the President, describing him as a soldier of India who always remained at the forefront of the battlefield and a comrade who has faced without getting rid of all the crises and troubles this country faced during the past generation. Sardar Patel also pledged unreserved loyalty and complete co-operation to the President in the 'heavy task of crossing the stormy seas that we have to face in the future.' He added that by his loving nature and goodness of heart, Dr. Rajendra Prasad had fully deserved the honour."

First Presidential Election, 1952

S. N. Candidate	Votes Polled
1. Dr. Rajendra Prasad	5,07,400
2. Shri K. T. Shah	92,827
3. Shri T. L. Ganesh	2,672
4. Shri Hari Ram	1,954
5. Shri K. K. Chatterjee	533
Total	*6,05,386*

Dr. Rajendra Prasad was declared as elected and the notice announcing this was published on May 06, 1952. Dr. Rajendra Prasad took up the office of the President of India on May 13, 1952.

Second Presidential Election, 1957

SI. No candidate	Votes Polled
1. Dr. Rajendra Prasad	4,59,698
2. Shri N. N. Das	2,000
3. Chowdhry Hari Ram	2,672
Total	**4,64,370**

Dr. Rajendra Prasad was declared as elected for a second term and the notice announcing this was published on May 10, 1957. Dr. Rajendra Prasad took up the office of the President of India on May 13, 1957.

Total Term of Office: January 26, 1950 to May 13, 1962

Chapter 11
The Ceremonial Procession

The Republic Day celebrations of 1950 in Delhi marked a grand procession. India was declared the Republic and Dr. Rajendra Prasad was its first President. Enjoyment filled all over and the President drove into the state. The colourful parade at Irwin Stadium followed him where the President hoisted the National Flag and took the Republic salute. This will remain in people's memory for a long.

It was people's day and they left no one in doubt about it. They crowded the streets, roofs, and available suitable places to see the Irwin Stadium ceremony. Even though they could not watch the President, the Prime Minister and the Cabinet Ministers being sworn in, they associated themselves with the ceremony by rushing into the open space of Government House and expressing their joy at the inauguration of the Republic.

While crowds were awaiting, the President in a white cap, black *achkan* and white *pyjamas*, stepped down the Darbar Hall steps to the State coach. The 35-year old coach was prepared for the occasion, bearing the new State emblem of Ashoka's Capital and drawn by six Australian Walrus (horse). As Dr. Rajendra Prasad stepped into the coach the joyful crowd rushed through the surrounded line.

The Preamble reads as follows;

WE, THE PEOPLE of India, having solemnly resolved to constitute India into a SOVEREIGN SOCIALIST SECULAR DEMOCRATIC REPUBLIC and to secure to all its citizens:

JUSTICE, social, economic and political; LIBERTY of thought, expression, belief, faith and worship;

EQUALITY of status and opportunity; and to promote among them all FRATERNITY assuring the dignity of the individual and the unity and integrity of the nation;

IN OUR CONSTITUENT ASSEMBLY this twenty-sixth day of November 1949, do HEREBY ADOPT, ENACT AND GIVE TO OURSELVES THIS CONSTITUTION.

Chapter 12
The Salient Features of the Constitution

India got independence in 1947 and was declared a republic on January 26, 1950. An active team of men and women - filled with patriotism, vision, and wisdom - spent over two years writing the distinctive Constitution, an important document because it governs the world's largest democracy.

Highlights:

The demand for the right of Indians to frame their own Constitution was made as early as 1895 - the Constitution of India Bill (1895), under the leadership of Lokmanya Tilak. It was the first non-official try. The Constituent Assembly of India held its first sitting on December 9, 1946:

It is comprised of 296 members. Of the 296, 15 were women and others men.

The Constitutions of the United States of America, Britain, Australia, Ireland and Canada served as major points of reference for the founders of our Constitution.

There was no adult franchise then and provinces had casual elections.

The Socialists, as a party, decided to avoid the Assembly. The Congress Party held 210 seats.

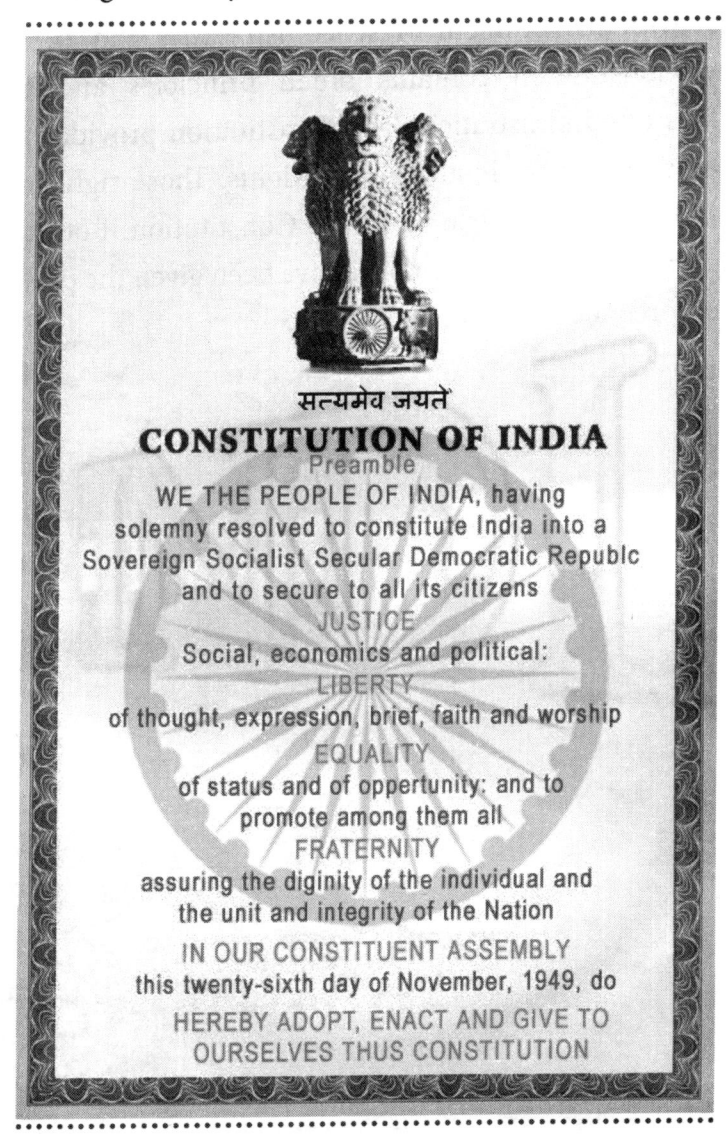

CONSTITUTION OF INDIA
Preamble
WE THE PEOPLE OF INDIA, having solemny resolved to constitute India into a Sovereign Socialist Secular Democratic Republic and to secure to all its citizens
JUSTICE
Social, economics and political:
LIBERTY
of thought, expression, brief, faith and worship
EQUALITY
of status and of oppertunity; and to promote among them all
FRATERNITY
assuring the diginity of the individual and the unit and integrity of the Nation
IN OUR CONSTITUENT ASSEMBLY
this twenty-sixth day of November, 1949, do HEREBY ADOPT, ENACT AND GIVE TO OURSELVES THUS CONSTITUTION

Dr. Babasaheb Ambedkar, who criticised the Congress Party throughout, was the Chairman of the Assembly.

The Constitution is probably the lengthiest in the world. It has 395 articles and ten schedules. 13 articles deal with the problem of scheduled castes and tribes. The Constitution contains broad principles and the details of administration. The Constitution provides for a large number of Fundamental Rights. Those rights are found in Articles 12 to 35 of the Constitution. Both the Supreme court and High court have been given the power to enforce the Fundamental Rights.

Chapter 13
Rajendra Prasad's Speech

The following is the speech made by Dr. Rajendra Prasad, President of the Constituent Assembly, on November 26, 1949, before passing Dr. Ambedkar's Constitution:

"I desire to congratulate the Assembly on accomplishing a task of such tremendous magnitude. Suppose you consider the population with which the Assembly has to deal. In that case, you will find that it is more than the population of the whole of Europe minus Russia, being 319 million as against 317 million. Apart from the size, other difficulties were inherent in the problem itself. We have got many communities living in this country. We have got many languages prevalent in different parts of it.

Another problem of great magnitude was the problem of the Indian States. When the British came to India, they did not conquer the country as a whole or at one stroke. They got bits of it from time to time. The bits, which came into their direct possession and control, became British India. Still, a considerable portion remained under the rule and control of the Indian Princes. The British thought that it was not necessary or profitable for them to take direct control of those territories. They allowed the old Rulers to continue, subject to their suzerainty. But they entered into various kinds of treaties and engagements with them. We had something near six hundred states, covering more than one-third of the territory of India and one-fourth of the country's population.

When the British decided to leave this country. The entire Paramountcy they had exercised so long and by which they could keep the Princes in order also lapsed. The Indian government was faced with the problem of tackling these states, which had different traditions of rule. Over time, not only have all the small states coalesced and become integrated with some province or other of India, but some of the larger ones also have joined..."

To the People of India

The following words of Dr. Rajendra Prasad are his message to the people of India:

"The method which the Constituent Assembly adopted in connection with the Constitution of India was first to lay down its 'terms of reference' as it were in the form of an Objective Resolution which was moved by Pandit Jawaharlal Nehru in an inspiring speech and which constitutes now the Preamble to our Constitution. It then appointed several committees to deal with different aspects of the constitutional problem. Several of these had as their Chairman, either Pandit Jawaharlal Nehru or Sardar Vallabhbhai Patel, to whom thus goes the credit for the fundamentals of our Constitution. I have to add that they all worked in a business-like manner and produced reports, which the Assembly considered. Their recommendations were adopted as the basis on which the draft of the Constitution had to be prepared. The Assembly then appointed the Drafting Committee, which worked on the original draft prepared by Mr. B. N. Rau and produced the Draft Constitution, which was considered by the Assembly at great length, at the second reading stage. As Dr. Ambedkar pointed out, there were not less than 7,635 amendments, of which 2,473 amendments were moved. We have now to consider the salient features of the Constitution. The first and the most obvious fact, which attracts any observer, is that we will have a Republic. India knew republics in the olden days, but that was 2,000 years ago or more and those republics were small republics. We have never had an elected Head of the State, which covered such a large area of India. And it is for the first time, it becomes open to the humblest and the lowliest citizens of the country to deserve and become the President or the Head of this big State, which counts among the world's biggest states today. This is not a small matter. All the powers rest in the Legislature, to which the Ministers are responsible.

Then, we come to the Ministers. They are, of course responsible to the Legislature and tender advice to the President who is bound to act according to that advice.

We have provided for adult suffrage by which the legislative assemblies in the Provinces and the House of the People in the Centre will be elected. It is a very big step that we have taken. It is big not only because our present electorate is a much smaller electorate and based very largely on a property qualification, but also because it involves tremendous numbers. We shall have not less than 160 million voters on our rolls. The work of organising elections by such vast numbers is of tremendous magnitude and there is no other country where an election on such a large scale has ever yet been held. Some people have doubted the wisdom of the adult franchise. I am not dismayed by it. I am a man of the village. I, therefore, know the village people who will constitute the bulk of this vast electorate. In my opinion, our people possess intelligence and common sense. They are not literate, but I do not doubt that they can measure their interest and the country's interests at large. I have therefore, no misgivings about the future on their account."

Chapter 14
His Respect for the Father of the Nation

The following passages are taken from his speech delivered by Dr. Rajendra Prasad while unveiling the statue of Mahatma Gandhi at Delhi.

True Happiness (October 11 1954)

It seems to have been taken for granted that we can raise the standard of living of human beings by requiring certain material resources. Following this principle, all the world countries are set upon acquiring and multiplying their resources. It is no doubt right that a hungry man cannot think of praying. Mahatma Gandhi himself once said that the hungry man sees God only in the form of bread. But even then, we should think about how far this kind of material prosperity can lead to real happiness.

I have also heard that the countries known to be prosperous and resourceful are not blessed with mental peace, whereas, on the other hand, we find lots of poor people, who excite our pity, leading a happy and contented existence. The truth is that the source of real happiness is in one's inner self and not in the outside world. We equate happiness with the world of external things and that is why, there is a scramble for acquisition and accumulation of things. The fact is that these things are, at best, no more than the means to achieve happiness and not happiness itself. One can experience happiness even without them. Apart from this, it is worthwhile considering what is real happiness.

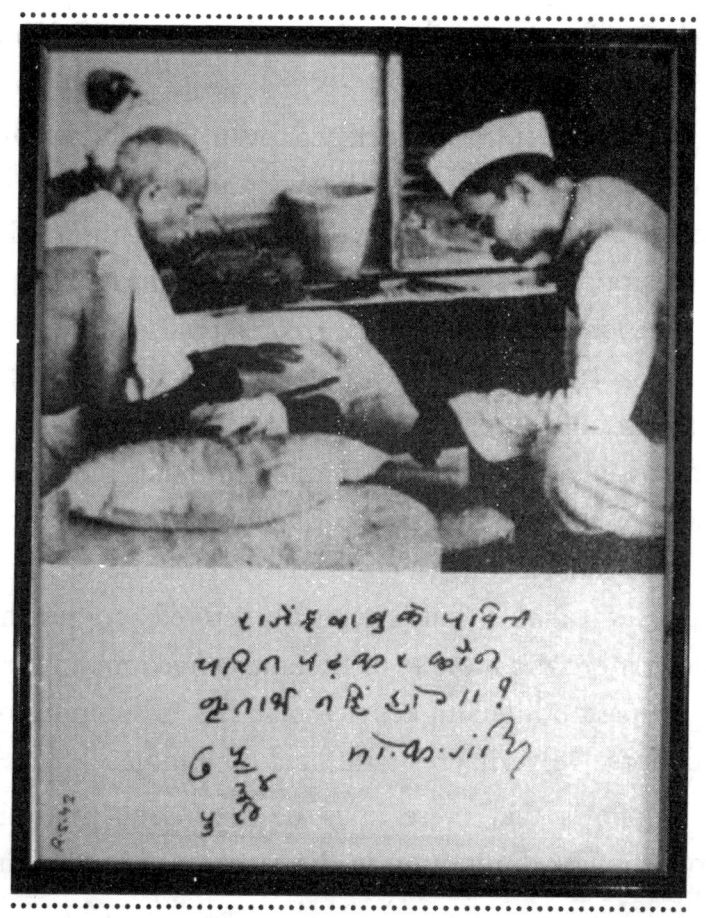

Real happiness or peace of mind means complete freedom from extraneous pressure or restraint or inhibitions. One basic fact, which must be recognised, is that any kind of inhibition or restraint is irksome. It ceases to be irksome only when it becomes something voluntarily accepted or adopted without restraint or coercion. This voluntary adoption of any line of thought or action, without external restraint or coercion, brings real happiness. Any subtraction from complete freedom is a loss of freedom to that extent and implies dependence on something else.

Man as a member of society or even an individual has long ceased to be fully free, if he ever was or can be free. All that can be aimed at or achieved is reducing or minimising this restraint or coercion and increasing to the maximum, the freedom that man enjoys. His material requirements can be satisfied, it is obvious, only by subjecting himself to some curtailment of this freedom. His mental satisfaction and possibly his spiritual aspiration becomes reduced in quantum and perhaps also in quality, by the amount of material satisfaction which in the very nature of things implies restraint. What is generally termed progress has tended more and more to restrict man's freedom. In every department of life and activity, man has to submit more and more to external restraints and inhibitions.

It follows that there must be a consequential and proportionate diminution in the mental satisfaction and spiritual endeavour, even though man may not feel that restraint or realise the evergrowing restraint being put on him from day today. It is thus clear that real happiness lies in freedom from restraint, which in turn, implies man's capacity to carry on with as little dependence on others as possible. We cannot escape from the conclusion that what is generally called a high standard of living has served to increase our dependence on others and to that extent has removed us further from real happiness.

Chapter 15
His View on India at the Time of Independence

"We see in the world of today that distance between country and country has almost been eliminated and nations living far apart from one another have come closer, so that if something happens at one place, it has its repercussion far and wide. It does not hold good about only dreadful things like war but also beneficent activities. One of the results of this progress has been that man is now dependent on his daily necessities in far-off countries. An example will clarify the point. Many of us present here today had known the days when the railway system in India was not expanded to the present extent, when there were no automobiles of any kind and when we had not even heard of the aeroplanes. At that time also, food was as important as it is today. Then, every community depended on its food on itself and on the land, which is cultivated. True, if a crop fails because of natural calamities like floods or drought, the community suffers. But otherwise, it managed to live on what it produced and learnt over time, the wisdom and the prudence to save food for emergencies. Regarding improving transport today, food grains can be easily supplied from one part of the country to another. We recently saw that food had to be dropped by aeroplanes on areas rendered inaccessible by the flood. All this sounds nice, but we have to see whether these developments have enhanced or restricted our freedom. My feeling is that by increasing such needs, as he cannot fulfil himself, man has necessarily restricted his freedom. By giving the example of food imports, I have tried to show our dependence on other countries. That is not all. If far off Argentina, Canada or America has a bumper wheat crop, India's falling wheat prices. Because of the improved means of transport, the availability, or otherwise, of things does not depend on local conditions but the overall

world conditions. If food cannot be imported from other countries because of natural calamity or war, the needy country will have to suffer untold misery. During the last war, we saw how even people of neutral countries had to suffer because of the restrictions on the export and import of certain articles from overseas. So, there are two aspects of this, progress. One promises plenty during peacetime. The other threatens to release a rich harvest of sufferings and privations, in case communications are dislocated on account of hostilities."

Chapter 16
His Thoughts on Freedom

"It is necessary to remember that even if all of our requirements are satisfied, we are selling our freedom for that satisfaction. For instance, whenever there is a disease in an epidemic form in the country, we must depend on other countries to provide medicines. Similarly, whenever there is a famine, others can save us from its dire consequences, but at the same time, if they like, they can also starve us by withholding the supply of food grains. If war breaks out today, the belligerents need not resort to deadly weapons to kill others. They can do it equally effectively by disrupting the system of transport. Therefore, while on the one hand, we are endeavouring to raise the standard of living, those very efforts might result in the curtailing of our freedom and independence.

Despite this overall progress, we have not yet reached a stage when we could produce an article in sufficient quantity to meet the requirements of all the people of the world. When we cannot say this about food, which tops the list of man's needs, it is no use talking about other things produced in still lesser quantities. That is why all the countries' living standards are not uniformly high and presents an unpleasant contrast. Those who possess more are anxious to extort more and more from those who do not possess much. The result is naturally conflicting between man and man and country and country. The fear of this conflict has become a nightmare for the modern man.

It is, therefore, necessary to realise that what we have assumed as axiomatic truth, namely that an increase in material prosperity also means the attainment of happiness, is neither quite correct nor so self-evident. This assumption is true only up to a certain limit and the more we transgress this limit, the more remote our chances of being happy become. This limit has to be fixed by the man himself. This is undoubtedly beset with countless difficulties, but I do think that it is not altogether impossible for man to achieve happiness without the usual paraphernalia, which passes for his everyday necessities. This is exactly what the adage means, 'simple living and high thinking. By practising this truth, Mahatma Gandhi could enjoy that happiness that a humble follower of his cannot even in the palatial Rashtrapati Bhavan.

I do not suggest that ambition, high aspirations, or desire for progress should be discouraged. But let us be sure that our will to progress and rise high will materialise in the true sense, only after we have realised that the source of our happiness does not lie outside us but is enshrined within our hearts. Our happiness will vary directly in proportion to the degree of our faith in the above truth. The more we try to achieve satisfaction, basing it on the outside world, the more we shall be inviting conflicts and depriving others of their joy."

Chapter 17
Words for Dr. Rajendra Prasad

By Mulayam Singh Yadav (Lucknow-December 3, 2003):

He asked the people of India to follow the principle that Rajendra Prasad followed to rule the country peacefully. He also said that Indians lack moral values that Gandhiji taught and Rajendra Prasad followed. He also talked about the life struggle of Rajendra Prasad.

(Speech given at 119th Jayanti at Rajendra Nagar Park)

A speech by Dr. APJ Abdul Kalam, the 11th President of India in his honour:

"There is no doubt that, beginning with Dr. Rajendra Prasad, the first President and the only one so far to have two terms, all heads of state of this country have been veteran politicians who tended to the presidency after long innings as Ministers, Chief Ministers, Governors and so on. Dr. S. Radhakrishnan may not have been a professional politician. Still, his experience in public affairs was enormous, including a spectacular stint as Ambassador to the Soviet Union in the Stalin era. Though primarily an educationist, Dr. Zakir Husain was deeply involved in national politics and the Freedom Struggle closely associated with Mahatma Gandhi."

Chapter 18
Birth of National Flag

The Indian flag, a tricolour, also called Tiranga was born on July 22, 1947, in the Constituent Assembly on the eve of the Independence of India. The Committee allotted for the flag adopted the National Flag of free India. Pandit Jawaharlal Nehru made a memorable speech and concluded by saying, "Sir, now I present to you not only the members but also the Flag itself".

The tricolour is a dedication to the sacrifices and blessings of all great souls, who brought freedom to India. It was a late evening of August 14, 1947 at 10:45 p.m. and the Central Hall of the Council Hall, now known as the Parliament House, was over packed to its capacity. At the given hour, the proceedings of the house commenced with the singing of Vande Matram led by Mrs. Sucheta Kriplani, the wife of then Congress President, Acharya Kriplani. A formal speech followed this, Dr. Rajendra Prasad, followed by Pandit Jawaharlal Nehru's famous speech, "Tryst with Destiny."

Finally, the members were moved to take the Oath of the Dedication. The text of the oath was:

"At this solemn moment when the people of India, by their suffering and sacrifice, have secured freedom and become martyrs of their destiny, I... a member of Constituent Assembly of India, do dedicate myself to the service of India and her people to the end that this ancient land attains its rightful and honoured place in the world and make its full willing contribution to the promotion of the world peace and welfare of mankind."

All the members took the oath standing. Its Chairman, Dr. Rajendra Prasad, read the oath, first in Hindi and then English. At that moment, every stone of the Parliament House echoed with the lusty shouts of *'Mahatma Gandhi ki jai'* and *'Vande Matram'*. After the oath-taking by the

House, Mrs. Hansa Mehta presented to the Chairman the 'Tiranga' on behalf of the women of India, symbolising the birth of the Indian National Flag. While presenting the flag to Dr. Rajendra Prasad, Mrs. Mehta said, "It is in the fitness of things that the first flag, that is to fly over this August House should be a gift from the Women of India."

Dr. Rajendra Prasad fondly received the flag from Mrs. Hansa Mehta and showed it around.

With the flag showing position of the Chairman, the proceedings of the historic day came to a close with the singing of *Sare Jahan Se Achcha Hindustan Hamara* and *Jana Gana Mana* (till then the song had not been adopted as the National Anthem of India).

On August 15, 1947, the dawn of Independence Day began at 8:30 a.m., with the swearing-in ceremony at the Rashtrapati Bhawan. The new government was sworn in the Darbar Hall. Two large national flags and the Governor General's flag in deep blue with the Star of India were hung in the backdrop on the hall wall facing the important gatherings.

The 'Tiranga' proudly went up for the first time against a free sky of Independent India, on the flag mast of the Council House at 10:30 a.m. As the tricolour went up the flag mast, a 31 gun salute was given, becoming the new nation's symbol. In the afternoon of August 15, 1947, the first public flag salutation ceremony was held at the War Memorial at the Prince's Park near India Gate. As the first Prime Minister of India unrolled the 'Tiranga' into the clear warm sky, a rainbow appeared suddenly on the horizon to bless 'Tiranga'. Lord Louis Mountbatten, the first Governor-General of free India, in his 17th Report, dated 16, 1947, wrote that the three colours, i.e. saffron, white and green in the flag, resembled the rainbow's colours. The Indian people told the whole event as a salute of Lord Indra, the God of Rains to the Tiranga.

Mera Bharat Mahan

'Tiranga' was hoisted for the first time on the wall of the Red Fort on the morning of August 16, which was a Saturday, at 8:30 a.m. and not on August 15 1947, as is commonly believed.

From 1948 onwards, the flag hoisting ceremony at the Red Fort was done on August 15. On that day, Pandit Nehru mentioned Subhash Chandra Bose's dream of seeing the National Flag hoisted on the Red Fort and regretted that he was not there to witness the day.

Chapter 19
The Indian Legislature

The Parliament consists of two groups of lawmakers, the Lok Sabha (House of the People—the lower house) and the Rajya Sabha (Council of States—the upper house). The Parliament's principal function is to pass laws on those matters that the Constitution specifies to be within its power. Among its constitutional powers are approval and removal of members of the Council of Ministers, amendment of the Constitution, approval of central government finances and lining up of state and union territory boundaries.

The President of India has a specific authority, concerning the function of the legislative branch. The President is given the power to call Parliament and accept all parliamentary bills before they become law. The President is empowered to summon Parliament to meet and address either house or both houses together and require the attendance of all of its members. The President also may send messages to either house regarding a pending bill or any other matter. The President addresses the first session of Parliament each year and must accept all provisions in bills passed.

According to its Constitution, India is a 'sovereign, socialist, secular, democratic republic'. India has a federal form of government. However, the central government in India has greater power in relation to its states and its central government is patterned after the British parliamentary system. The government exercises its broad authority powers in the name of the President, whose duties are largely ceremonial. The President and Vice President are elected indirectly for 5-year terms by specially elected members. After the five-year term comes to an end the President may resign from his office by writing to the Vice-President of India. But, he will continue to hold his office, until his successor takes up his office. The Vice-President acts as his substitute if President's office falls vacant on the grounds of his death, resignation or impeachment or otherwise. A vacancy should be filled by an election necessarily taking place within six months of his office falling vacant.

Real national executive power is centred in the Council of Ministers (cabinet), led by the Prime Minister. The President appoints the Prime Minister, appointed by lawmakers of the political party or combining commanding a parliamentary majority. The President then appoints subordinate ministers, on the advice of the Prime Minister.

India's Parliament consists of the Rajya Sabha (Council of States) and the Lok Sabha (House of the People). The Council of Ministers is responsible to the Lok Sabha.

The Legislatures of the States and Union Territories elect 233 members to the Rajya Sabha and the President appoints another 12. The elected members of the Rajya Sabha serve for a 6-year term, with one-third up for election every two years. The Lok Sabha consists of 545 members, 543 are directly elected to 5-year terms. The other two are appointed.

India's independent judicial system began under the British and its concepts and procedures are the same like those of the Anglo-Saxon countries. The Supreme Court consists of a Chief Justice and 25 other Justices, all appointed by the President on the advice of the Prime Minister.

India has 28 states and 8 union territories. At the state level, some of the legislatures are bicameral, patterned after the two houses of the national Parliament. The States' Chief Ministers are responsible to the legislatures, in the same way as the Prime Minister is responsible to Parliament.

Each state also has a presidentially appointed Governor who may assume certain broad powers directed by the central government. Although some territories have gained more power to administer their affairs, the central government exerts greater control over the Union

Territories than the States. Local governments in India have less independence than the United States. Some states are trying to make the traditional Village Councils or Panchayats, aiming to promote popular democratic participation at the village level, where much of the population still lives.

Chapter 20
The President House

The appearance of the Rashtrapati Bhavan is multi dimensional. It is a big mansion and its architecture is excellent. More than these, it has a hallowed existence in the annals of democracy, for being the residence of the President of the largest democracy in the world. Few official residential premises of the Head of the State in the world will match the Rashtrapati Bhavan in its size, vastness, and magnificence.

The present-day Rashtrapati Bhavan was the residence of the British Viceroy. It was Sir Edwin Landseer Lutyens who conceptualised the architecture of this building. The decision to build a residence in New Delhi for the British Viceroy was taken after it was decided in the Delhi Darbar of 1911 that the capital of India would be shifted from Calcutta to Delhi in the same year. It was constructed to show the permanence of British rule in India.

Rajendra Prasad became the first President of India and occupied this building with preserving, protecting, and defending India's Constitution. From that day, this building was renamed as Rashtrapati Bhavan - the President's House.

It is interesting to note that the building, which was scheduled to be completed in four years, took seventeen years and on the eighteenth year of its completion, India became independent.

This big mansion has got four floors and 340 rooms. With a floor area of 200,000 square feet, it uses 700 million bricks and three million cubic feet of stone. Hardly any steel has gone into the construction of the building.

Chapter 21
Some Quotes on Dr. Rajendra Prasad

- Sarojini Naidu, "He was to Gandhiji, what John was to Christ."
- Jawaharlal Nehru, "He is the symbol of Bharat."
- Gaṇdhiji, "There is at least one man who would not hesitate to take the cup of poison from my hands."
- Gunter, "heart of the congress organisation."
- Dr. Rajendra Prasad's postal stamp of denomination 15np was released on May 13, 1962 by the Indian Posts & Telegraph Department. On this occasion, the Indian Posts & Telegraph Department also mentioned that "In issuing a special commemorative stamp in honour of Dr. Rajendra Prasad, the Posts and Telegraphs Department of the Government of India consider themselves privileged in being able to offer a token of the country's immense gratitude to this illustrious son of India.